Biotechnology in Agriculture, Industry and Medicine Series

Perspectives on Lipase Enzyme Technology

Biotechnology in Agriculture, Industry and Medicine Series

Agricultural Biotechnology: An Economic Perspective
Margriet F. Caswell, Keith O. Fuglie, and Cassandra A. Klotz
2003. ISBN: 1-59033-624-0

Biotechnology in Agriculture and the Food Industry
G.E. Zaikov (Editor)
2004. ISBN: 1-59454-119-1

Governing Risk in the 21st Century: Lessons from the World of Biotechnology
Peter W.B. Phillips (Editor)
2006. ISBN: 1-59454-818-8

Biotechnology and Industry
G.E. Zaikov (Editor)
2007. ISBN: 1-59454-116-7

Research Progress in Biotechnology
G.E. Zaikov (Editor)
2008. ISBN: 978-1-60456-000-8

Biotechnology and Bioengineering
William G. Flynne (Editor)
2008. ISBN: 978-1-60456-067-1

Biotechnology: Research, Technology and Applications
Felix W. Richter (Editor)
2008. ISBN: 978-1-60456-901-8

Biotechnology, Biodegradation, Water and Foodstuffs
G.E. Zaikov and Larisa Petrivna Krylova (Editors)
2009. ISBN: 978-1-60692-097-8

Industrial Biotechnology
Shara L. Aranoff, Daniel R. Pearson, Deanna Tanner Okun, Irving A. Williamson, Dean A. Pinkert, Robert A. Rogowsky, and Karen Laney-Cummings
2009. ISBN: 978-1-60692-256-9

Industrial Biotechnology: Patenting Trends and Innovation
Katherine Linton, Philip Stone, Jeremy Wise, Alexander Bamiagis, Shannon Gaffney, Elizabeth Nesbitt, Matthew Potts, Robert Feinberg, Laura Polly, Sharon Greenfield, Monica Reed, Wanda Tolson, and Karen Laney-Cummings
2009. ISBN: 978-1-60741-032-4

Biosensors Properties, Materials and Applications
Rafael Comeaux and Pablo Novotny (Editors)
2009 ISBN: 978-1-60741-617-3

Perspectives on Lipase Enzyme Technology
J. Geraldine Sandana Mala And Satoru Takeuchi
2009 ASBN: 978-1-61741-977-8

Biotechnology in Agriculture, Industry and Medicine Series

PERSPECTIVES ON LIPASE ENZYME TECHNOLOGY

J. GERALDINE SANDANA MALA
AND
SATORU TAKEUCHI

Nova Science Publishers, Inc.
New York

For permission to use material from this book please contact us:
Telephone 631-231-7269; Fax 631-231-8175
Web Site: http://www.novapublishers.com

LIBRARY OF CONGRESS CATALOGING-IN-PUBLICATION DATA

Mala, J. Geraldine Sandana.
Perspectives on lipase enzyme technology / J. Geraldine Sandana Mala, Satoru Takeuchi.
p. ; cm.
Includes bibliographical references and index.
ISBN 978-1-60741-977-8 (hardcover)
1. Lipase--Biotechnology. I. Takeuchi, Satoru. II. Title.
[DNLM: 1. Lipase--metabolism. 2. Biotechnology. 3. Lipase--isolation & purification. QU 136 M236p 2009]
TP248.65.E59M35 2009
660.6'34--dc22

2009025843

Published by Nova Science Publishers, Inc. ✝ New York

Dedicated to My Parents

— J. Geraldine Sandana Mala

CONTENTS

PREFACE

Perspectives on Lipase Enzyme Technology presents a discussion of a sub-discipline in biotechnology suitable for biologists at various levels and is designed to provide a comprehensive review on lipases, in particular from the microbial sources, thereupon to impart good scientific perceptions to the reader in this discipline. It is proposed to reach the scientific community, especially research fellows at the start of their research investigations in pursuit of planning a research strategy.

This book also provides an interface between biochemistry and microbiology graduates to synergize into the biotechnology streamline and is designed to provide a comprehensive approach in terms of its suitability for young and experienced researchers. Experimental protocols are presented for easier laboratory practices and it is hoped that this book would be a good reference for the reader in this area of research, in particular in developing a research strategy for investigations right from isolation and production methods of microbial lipases to its applications in the industrial scenario. This book would probably take a different approach in the context of elaborate discussions in a practical perspective. It includes a section on bioinformatics.

The authors first describe the science of lipases with an introduction on the versatility of their catalysis and specificities. The theoretical and practical aspects for isolation of suitable microorganisms for lipase production and their requirements for optimal production in suitable media are detailed. Lipase purification and its molecular cloning are important criteria for successful commercial exploitation, and are therefore explained in detail. A chapter in bioinformatics is included in view of its current thrust. Conclusions regarding the potentialities of lipases in terms of their profound applications as increasingly important biocatalysts are given.

The book is supplemented with experimental procedures and important information in the Appendix is discussed for ready reference. Coloured illustrations and technical information in boxes are provided to improve the quality of presentation. Overall, the book attempts to cater to the needs of ongoing research activities in this area of specialization.

ACKNOWLEDGMENTS

The authors gratefully acknowledge the financial support of Mr. Kikuji Takeuchi and Mr. Naomi Takeuchi of Takenen, Japan.

Dr. J. Geraldine Sandana Mala wishes to sincerely thank Dr. Satoru Takeuchi, President of Takenen, Japan, for his constant encouragement and support in rendering the completion of this book. She also expresses her deep sense of gratitude to her parents, Mr. A.J. John Bromeo and Mrs. Helen Bromeo; sister, Ms. J. Rita Jasmine Ranjani; and brothers, Mr. J. Maria Thomas and J. Justin Arul Xavier for their deep affection and moral support towards the preparation of this book. Illustrations provided by Ms. J. Rita Jasmine Ranjani and lipase structures from Google Images are also acknowledged.

Chapter 1

BIOTECHNOLOGY: THE SCIENCE AND ORIGIN OF LIPASE

1.1. INTRODUCTION

Biotechnology, in a broad sense, could be defined as an integrated approach of biochemistry, microbiology and genetic engineering and, in general, life sciences, targeted towards beneficial factors for mankind and socio-economics.

The United Nations Convention on Biological Diversity defines biotechnology as 'any technological application that uses biological systems, living organisms, or derivatives thereof, to make or modify products or processes for specific use'. Biotechnology is an emerging discipline of research interests since 1970s and has made a tremendous impact especially in healthcare, pharma and agri sectors, industrial developments and environmental concerns, providing a platform for 'next generation' modalities. In a wholesome consideration, biotechnology presents a clear basis of scientific knowledge in exploring and unraveling biological information. In essence, it acts as a nodal point synergizing the biology and technology interface.

The rapidity of biotechnology is conceptual from age-old brewery to recent stem cell therapy and genome biology. A remarkable success story is the discovery of the double helical structure of DNA by the historical pioneers, Francis Crick and James Watson, the basis of which had formed the genomic era.

Genetic engineering, and further, genomics, has progressed from the discovery of restriction enzymes which act as important cutting tools in molecular cloning to the development of microarray technology using *in silico* and high-throughput technologies.

One of the greatest milestones so far accomplished has been the HUMAN GENOME PROJECT, an initiative of Department of Energy (DOE) of USA since 1986 and officially begun in 1990. A joint effort of six major countries, the USA, UK, France, Germany, Japan and China enabled rapid announcement of completion of the genome sequencing by April 2003, quite earlier than expected with technological advancements of the sequencing procedures.

Box 1.1. Discovery of the Double Helix

The discovery of the double helix of DNA made a remarkable landmark in the history of Molecular Biology, which appeared in Nature, April 1953 by Francis Crick and James Watson who won their Nobel Prize in Physiology or Medicine in 1962 along with Maurice Wilkins. This was a result of a series of events at the Cavendish Laboratory in Cambridge and the Kings College in London. The efforts of DNA modeling, X-ray crystallography and discussions amongst prominent scientists led to the unraveling of DNA structure. Earlier, DNA was not thought to possess hereditary information or the 'blue print' as known today, due to complexity of protein structures. Later, upon identification of DNA as the genetic material, DNA was presumed to be a triple helix. It was then deduced that DNA consists of nucleotide bases, although specific base pairing was not proved. From X-ray crystallography, Rosalind Franklin noted the pairing of bases and postulated that the base pairs were inner while the phosphate was outer along the polynucleotide chain. This remarkable discovery enabled the scientists to confirm and identify the 'double helical' structure of DNA by comparisons with their DNA modeling analysis and crystallographic data.

The double helical structure of DNA laid the basis for DNA replication by the 'semi-conservative model' and consequently the Central Dogma of Molecular Biology which reveal the information flow from DNA to proteins by transcription and translation.

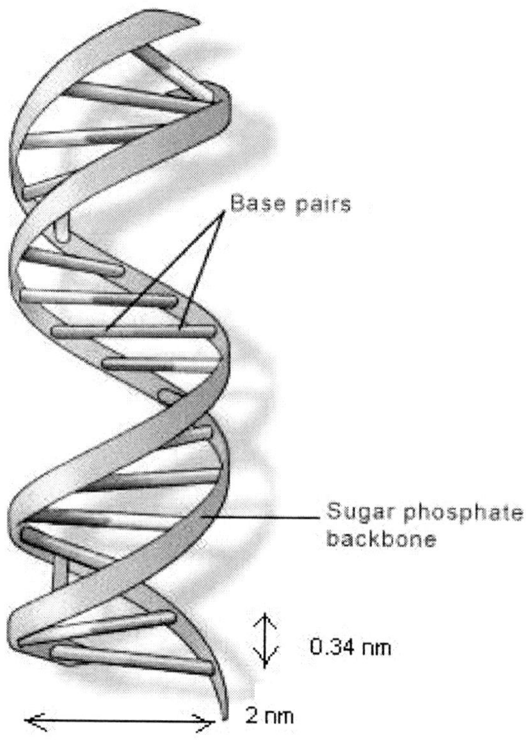

Box 1.2. Timeline of the Human Genome Project (1990–2003)

1988	NIH and DOE sign up for an MoU to coordinate human genome research activities.
1989	National Center for Human Genome Research is established for the NIH component. James Watson appointed as the first Director.
1990	A Five year plan and the National Advisory Council for the Human Genome Research are established. The Human Genome Project officially begins.
1991	National Advisory Council for Human Genome Research meet in MD.
1992	Michael Gottesman is appointed acting NCHGR Director.
1993	Francis Collins is appointed NCHGR Director. US HGP revises its Five year goals.
1994	Human Genetic Mapping goal achieved.
1995	The Task Force on Genetic Testing is established as a subgroup of the NIH/DOE Ethical, Legal and Social Implications Working Group.
1996	ELSI Report is issued by the Joint NIH/DOE Committee evaluating the Ethical, Legal and Social Implications program of the Human Genome Project.
1997	Human chromosome 7 map completed. NCHGR is renamed National Human Genome Research Institute.
1998	The HGP plans to generate a 'working draft' in 2001 that, together with the 'finished sequence' in 2003, will cover at least 90 percent of the genome.

A new five-year plan for the HGP for the next five fiscal years 1999–2003 is developed.

1999	National Human Genome Research Institute (NHGRI) and other HGP-funded scientists unravel for the first time the genetic code of an entire human chromosome.
2000	Scientists in Japan and Germany unravel the genetic code of human chromosome 21. The Human Genome Project is the recipient of the American Society of Human Genetics' Allan Award' to honor the hundreds of scientists involved in deciphering the human genetic code.
2001	International Human Genome Sequencing Consortium provides the first analysis of the human genome sequence that describes how it is organized and how it evolved. A significant discovery reveals that there are only 30,000-40,000 genes, not 100,000 as presumed earlier.
2002	NHGRI launches a new Website, *www.genome.gov*, that provides improved usability and easy access to new content
2003	In April, NHGRI celebrates the completion of human genome sequence ahead than drafted. The commemoration of the 50th anniversary of the discovery of DNA double helix is felicitated. The vision document for the future of genomics research is published.

Reverse transcriptase, discovered in 1970, has also increased the prospects of molecular techniques in genomic and viral research. Biotechnology has further evolved since the millennium in the applications of RNAi, embryonic and adult stem cells, transgenics of microbes, animals and crops and innumerous other measures and has futuristically increased the scope of today's Biotech world.

Biotechnology in medicine and agriculture being of immense research interests, biotechnologically relevant enzymology has fascinated researchers with invincible interests in aspects of Bioprocess Technology and Molecular Biotechnology. Convergence of

Biotechnology entity into industrially important enzymology for commercially viable products has surged keen interests in almost all areas of Life Sciences. The practice of Industrial Microbiology has been revolutionized by exploitation of microbes for commercial applications. One of the most striking features is the evidence of presence of microbes even in meteorites, which are potential candidates for enzyme production and easier genetic manipulations.

Biotechnologically relevant enzymes include the most studied proteases, carbohydrases and lipases. Lipases have been significantly exploited for its versatile applications due to multiplexity of catalysis and has prompted increasing Basic and Applied research since the past two decades. Lipases form the third largest group of industrial enzymes with significant biotechnological potentials. Several Companies have also resorted to lipase production to meet the mass requirements. Novo Nordisk, Denmark, and Amano Enzyme Inc., Japan, produce most of the commercial lipases due to enormous demand for lipases in various industrial sectors.

1.2. LIPASES: AN INTRODUCTION

Lipases are triacylglycerol acylhydrolases, E.C. 3.1.1.3, which catalyse the hydrolysis of triglycerides, commonly, fats with a glycerol backbone, to yield concomitant free fatty acids (FFA) and glycerol moieties.

The reaction is reversible, also to catalyse the esterification of FFA and glycerol in a nonaqueous medium. Lipases act at the oil-water interface requiring an insoluble substrate phase, lipid substrates. In general, lipases do not require any cofactor for its catalytic activity, unlike the cellular metabolic enzymes. Lipases are classified as 'true lipases' which act on long-chain fatty acid substrates, while the enzymes with short-chain fatty acid preferences are classified as 'esterases'. Most commonly, triglyceryl oleates (triolein) are the true substrates for lipase catalysis and the commercially available olive oil is also a much-suited substrate in the laboratory, which has more than 70% oleate residues and lesser traces of impurities that need not be considered for commercial applications. However purification methods, subsequent characterization data and structural analyses require the pure trioleins.

The multiplexity of lipase catalysis, involves hydrolysis, esterification, transesterification and interesterification reactions. These reactions promise a wide array of industrial applications.

Box 1.3. Enzyme Commission Nomenclature of Lipase

E.C. 3.	Hydrolase
E.C. 3.1.	Acting on Ester bonds
E.C. 3.1.1.	Carboxylic ester hydrolase
E.C. 3.1.1.3.	Triacylglycerol Lipase

Box 1.4. Lipase Catalysis

TAG	Hydrolysis	$RCOOR_1 + H_2O \leftrightarrow RCOOH + R_1OH$
↕		
DAG	Esterification	$RCOOH + R_1OH \leftrightarrow RCOOR_1 + H_2O$
↕		
MAG	Transesterification	$RCOOR_1 + R_2COOH\ (R_2OH) \leftrightarrow RCOOR_2 + R_1COOH\ (R_1OH)$
↕		
G + FFA	Interesterification	$RCOOR_1 + R_2COOR_3 \leftrightarrow RCOOR_3 + R_2COOR_1$

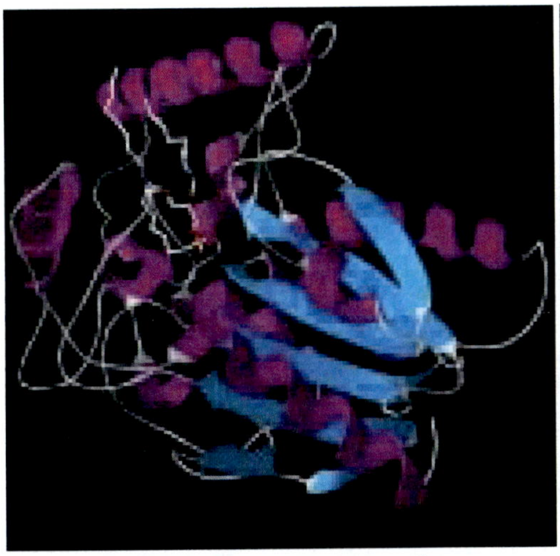

Pseudomonas aeruginosa lipase

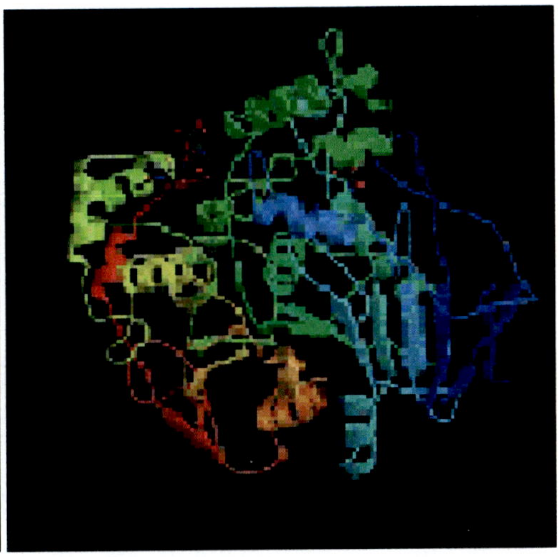

Candida rugosa lipase

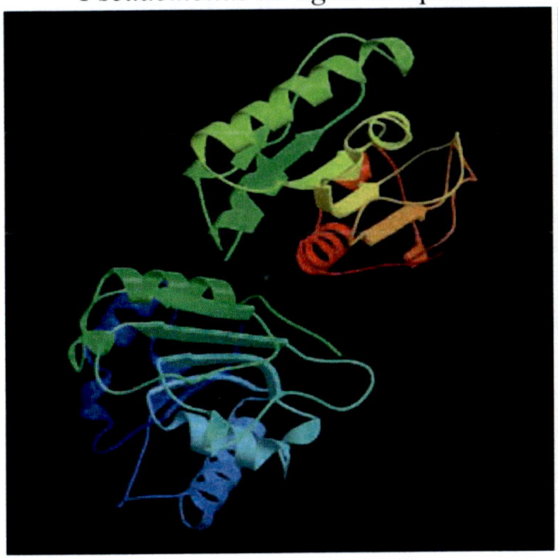

Bacillus subtilis lipase

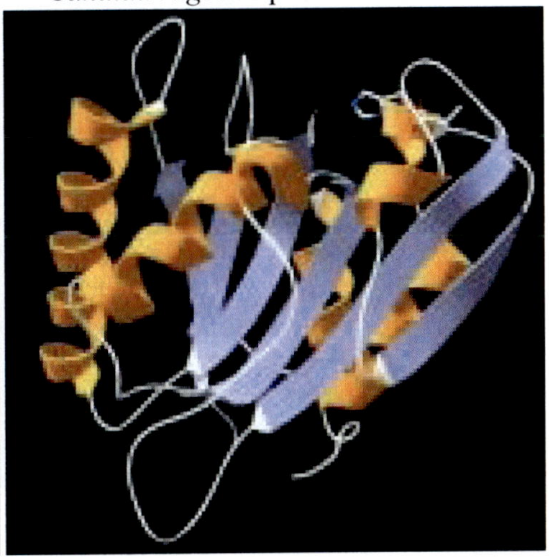

Bacillus subtilis A lipase

Figure 1.1. 3-D lipase structures of microbial origin.

1.3. Lipases: The Structure

Lipases are small molecular enzymes with 20–60 kDa MWs and are usually monomeric polypeptide chains. Oligomeric polypeptide chains have also been reported in few other investigations. All microbial lipases contain a consensus sequence of G-S-X-S-G in their protein architecture and a significant catalytic triad containing Ser-His-Asp similar to serine proteases. In case of *Geotrichum* sp., Asp may be replaced by Glu. An active serine is proposed to be present under a short helical fragment of a long surface loop. Arg residues play an important role in stabilization of the active open-lid conformation. Structural mechanisms so far interpreted provide a concise view, common to most lipases. The active site is buried beneath a helical segment or the 'lid' structure, which opens up upon contact with a substrate molecule. A large hydrophobic patch uncovers the lid and makes the active site accessible to the lipid substrate. This process accounts for inducing a favourable conformation of the enzyme molecule, known as 'interfacial activation', a unique property of the lipase family. The core of lipase (catalytic domain) is an alpha/beta-sheet, containing 8 strands connected by helices. This conserved structural feature of the fold is referred to as the 'α/β-hydrolase fold' common to hydrolytic enyzmes of different phylogenetic origin and catalytic function.

1.4. Lipases: Specificity

Lipases possess remarkable specificities amongst their own as well as different genera. The specificities of only microbial lipases are described in view of the scope of this book. Thereby, according to nature of the cleavage of ester bonds and yield of products, lipases are classified into major subgroups as: non-specific; 1,3-regiospecific; fatty acid specific; substrate specific and stereospecific enzymes. Non specificity is attributed to the cleavage of any of the three bonds on the glycerol moiety esterified with three fatty acid moieties, which yield diglycerides, monoglycerides and further free fatty acids upon longer hydrolysis. A classic example of this type of specificity is the lipase from *Candida rugosa.* In contrast, 1,3-regiospecific lipases are from *Aspergillus niger* which targets only the first and third positions or both to yield 1,2 (2,3) diglycerides and 2-monoglycerides. Fatty acid specificity is unique to the class of *Geotrichum candidum* lipases which hydrolyse at fatty acid esters of only unsaturated bonds, at *cis*-9 position of C 18:1 fatty acid. Substrate specificity of *Penicillium camembertii* lipase is remarkable for its catalysis of the hydrolysis of only diglycerides and monoglycerides, but not triglycerides. Stereospecificity is the ability to distinguish *sn*-1 or *sn*-3 positions of fatty acid ester bonds in triglycerides, in the case of lipase. This is observed in *Pseudomonas aeruginosa* which displays *sn*-1 stereospecificity.

Determination of the positional specificity of lipases is relatively simple to perform on silica gel TLC plates along with standard samples of relevant products of hydrolysis (co-TLC). This method has been further enhanced using radioactive substrates, which can be detected by autoradiography. Still further methods could be used by detecting with Rhodamine B, which forms fluorescent spots of the products. Another simpler method could be detection using Sudan fat stains.

1.5. LIPASES: KINETICS

Lipase kinetics is a debated study due to heterogeneous interface kinetics which does not follow the conventional Michaelis-Menton kinetics. The rate of the lipase reaction depends only on the concentration of enzyme-substrate within the surface phase. The partitioning of the enzyme from the aqueous phase to the surface phase is distinct from the interaction of the enzyme with its substrate within the surface phase. Thus, a typical environment of lipase function consists of a lipid droplet separated from the aqueous milieu by a two-dimensional surface phase.

1.6. LIPASES: ACTIVITY DETERMINATIONS

Lipase activity has to be determined for comparisons between lipases of different sources and estimations of their degrees of potency. Preliminary investigations are essential for qualitative and approximate quantitation of lipase activities for a simple and speedy approach. These methods listed below should be definitely followed by precise quantitation methods.

Box 1.5. Lipase Plate Assays

Tributyrin-agar Diffusion:	Tributyrin is used as the substrate and lipase activity is an estimate of the halozone produced.
Rhodamine B-agar Diffusion:	Fluorescent Rhodamine B is used to enhance visualization of the halozone of olive oil or tributyrin-agar plates.
$CaCl_2$-Tween 80 Diffusion:	Precipitation zone is observed in $CaCl_2$-Tween 80 plates due to production of calcium oleates of fatty acids.

Box 1.6. Definition of Lipase Activity

Lipase activity is defined as the amount of enzyme required to hydrolyse one μmol of FFA liberated per minute under standard assay conditions. It is expressed as Units/ml/min (μmol/ml/min).

$$\text{Lipase activity (by titrimetry)} = \frac{\text{Vol. of NaOH consumed (ml) x Normality of NaOH (N) x 1000}}{\text{Time of incubation (min) x Vol. of lipase added (ml)}}$$

1000 is a factor for converting mN into μN (for NaOH, N is same as M) = μM = μmol/ml
Volume of enzyme used is 1.0 ml
Time of incubation is 20 min.

Lipases possess significant substrate specificity, and hence, a number of assay protocols have been developed based on this property. Lipolytic activity is generally signified by the accumulation of FFAs. This is often a simpler measure of lipase activities using insoluble triglycerides of long-chain fatty acids. Thus, lipase assays have evolved primarily on this basis and performed under standard conditions. The assay conditions affect the lipase activity significantly and usually require proper standardizations in the laboratory. The simple assumptions of lipase assay procedures are that FFAs are liberated at 1:1 ratios of FFA and glycerol moieties. Hence, an increase in FFAs suggests a higher degree of lipase activity under standardized conditions.

To increase FFA detection, additives such as Ca salts and detergents maybe employed. Ca salts such as $CaCl_2$ prevent prolonged accumulation of FFAs which tend to inhibit enzyme activity by product inhibition. Formation of Ca salts of fatty acids clears away the FFAs in the reaction mixture. This prevents accumulation of FFAs at the enzyme-substrate interface. Detergents, on the other hand, increase the availability of substrate molecules at the interface thereby providing more accessibility of the enzyme towards its substrate for enhanced reactivity. However, the presence of additives may also have inhibitory effects on the lipase molecule due to unfavourable conformations or metabolic preferences. Physico-chemical factors also affect lipase reactivity, such as buffer pH, incubation time and temperature. Hence, a careful choice of physico-chemical factors and additives is necessitated for devising proper assay strategies. This type of assay protocol is the basis of titrimetric estimations which is acid-base titrimetry employing standardized alkali, usually NaOH, using an olive oil-polyvinyl alcohol emulsion as the substrate. Emulsification of olive oil is required due to its immiscibility in water, for access of the lipase enzyme to this substrate at the interface, whereby more substrate molecules could be encountered. Tributyrin is another substrate for titrimetry, due to its partial miscibility in water, and can be used for esterase activities upon homogenization, but is generally not recommended for estimations of true lipases. A rather alternate titrimetry than using conventional titration apparatus is the use of a pH-stat that defines a more accurate end point of the reaction system, set to alkaline pH values. Titrimetry is a preferred choice for lipase assay procedures used for commercial or lab-scale experiments that involve media standardizations upto the order of 10 and above lipase units that do not require accuracy of values.

More precisely, purification analyses require spectrophotometric estimations which use the absorbance of esters in the visible range due to coloured product components. An example of this is the use of p-nitrophenyl esters (p-nitrophenyl palmitate/laurate) that liberate p-nitrophenol which under alkaline conditions can be estimated at A_{410} nm. One molecule of p-nitrophenol is equivalent to one molecule of the FFA released. Thus, lipase activity can be expressed in terms of p-nitrophenol liberated by the enzyme. Several other chromogenic substrates have been widely used for spectrophotometric estimations of lipase activities. These include α- and β- napthyl esters, particularly for low temperature-active lipases used in dairy applications. Another spectrophotometric estimation is the turbidimetric method using Tween 20 as the substrate in which turbidity of the reaction mixture increases coordinately upon forming Ca salts and can be estimated at A_{500} nm. However, this method is out of common use due to discrepancies in measurement of turbidity due to other factors also.

A day-to-day laboratory estimation for lipase activity is the copper soap colorimetric assay developed in 1986. This protocol uses cupric acetate in the presence of pyridine (to adjust to pH 6.1) yields a colored complex upon lipase hydrolysis and the isooctane phase containing the fatty acid-copper complex is measured at A_{715}. Use of isooctane results in flexibility of the pH value of the copper reagent between pH 5.8-6.4 and read at A_{715}. This method is easier and simpler to perform with good accuracy compared to arbitrary values of the titrimetric procedures.

Other methods used for lipase activity analysis are GC, TLC and ELIZA techniques. Atomic force microscopy and IR spectroscopy are also other lipase assay techniques. Advancements in assay procedures such as Fluorescence, HPLC and Radiolabelling techniques yield better accuracies than conventional assays. However, the cost factor of lipase assays is also an important criterion for routine estimations. With developments in the computer age, subsequent sophistications of lipase determinations have resulted in the programming of software packages to streamline lipase activity evaluations of the different assay modules.

1.7. Lipases: Taxonomy

Lipolytic activity is a significant characteristic of any organism in view of metabolism of fats and fatty acids and is therefore of taxonomic importance of varied species, however exhibiting differences in their activities. In the case of microbial genera, lipase producers are widespread and may be from bacteria, fungi and yeasts. The potency and characteristics of lipases vary among different microorganisms. Taxonomic distribution is an important factor imparting unique features for the origin of the microbial sources.

Major genera of lipase producers are *Pseudomonas*, *Chromobacterium*, *Staphylococcus*, *Alcaligenes* (bacteria), *Penicillium*, *Geotrichum*, *Humicola*, *Rhizopus*, *Mucor* (fungi) and *Candida*, *Trichosporon*, *Yarrowia* (yeast).

Lipases of most microorganisms are inducible in nature. Constitutive lipases may also occur, but in rarity. Lipases are generally extracellular or in few cases, intracellular. Extracellular lipases are those in which lipase is secreted into the fermentation medium and do not require cell sonification procedures for their isolations, while intracellular lipases require separation methods from the microbial cells for isolation in their native forms. Multiple forms of lipases also exist as in the case of *Geotrichum candidum* lipase. This means that different lipase forms can be produced by a single species, but retain their unique characteristics. The lipase forms also require different purification methods and can be utilized for their differences in properties.

1.8. Lipases: Sources

Lipases are ubiquitous in nature and may be produced by plants, animals and microorganisms. Plant sources are rice bran, castor bean, wheat germ, sunflower and oat seeds. However, plant sources require tedious extraction procedures with only lesser yields. Centrifugation and chromatography techniques are necessitated and add to the complexity of isolation methods. Contact with potent allergenic proteins and dry seed powders may be harmful. Animal sources are pancreatic, hepatic, or blood. Pancreatic lipases have been extensively studied and require cofactors for the stimulation of its activity. Also, structural features of pancreatic lipases have been detailed. In terms of isolation methods, pancreatic lipase needs native conditions and is also tedious.

Microorganisms are the most preferred sources due to their potencies and ease of production. Lipases from microbial sources could be scaled up for increased productivities by a source of single culture and could also be overproduced by genetic manipulations for desired characteristics. Microbial sources are *Candida* sp., *Pseudomonas* sp., *Rhizopus* sp. to name significant producers of yeast, bacteria and fungal origins.

1.9. Lipases: Applications

Lipases exhibit multiplexity of catalysis in aqueous as well as organic media which route different avenues for their profound industrial applications. As described earlier, lipases possess remarkable properties that make them ideal candidates in many of the industrial sectors. The specific characteristics of lipases that enable their wide applicability is their catalyses of reactions destined to produce useful reaction products of significant commercial values, which also form the raw materials for various other commercial applications. Apart from this their criteria of specificities also render them more suitable for commercial exploitation with desired choice of reactions products. The industrial applications of lipases have been exploited in the past few decades with a recent surge in basic as well as applied research. Consequently, the industrial importance of these enzymes implicates the necessity to suit viability and commercial productivity. Lipases have been conventionally used in food, dairy, pharmaceutical, medical, oleoprocessing, detergent, leather and several other allied sectors. These industrial components have been indicated in Box.1.7.

Box. 1.7. Applications of Lipases

Food	fragrance and flavours, meat processing
Dairy	cheese ripening, coffee whiteners, confectionery
Pharmaceutics	resolution of racemic mixtures for isolation of biologically active compounds
Medical	heart ailments, biosensors, treatment of malignant tumors
Oleoprocessing	hydrolysis of oils and fats, synthesis of esters, interesterification/ transesterification of oils and fats
Detergent	formulations as additives in detergents for fat/grease removal of fabrics
Leather	degreasing of hides and skins to prevent fatty spues in finished products

Conclusion

Lipases are therefore renowned for its scientific basis of commercial exploitations. Lipase enzyme technology thereby encompasses a wide range of analytical as well as industrial potentials in areas such as pharmaceutics, medicine and oleochemistry with versatility of catalysis and increasing attention in basic research for structural evaluations. Protein engineering of lipases have been of research interests since the mid-1980s based on sequence information of *Ps. mendocina* and increased dramatically for other lipase structures also. However, this field is still in its infancy due to complexity of catalysis, insoluble lipid substrates, and few ideas about targeted applications. In future, lipase research could also be focused towards structure-function analysis to establish the basic mechanisms underlying the multipotent features for its desired characteristics and adaptability for protein engineering and therapeutic measures.

Self-evaluation

(1) Human Genome Project was initiated in
(a) 1990 (b) 1993 (c) 1996 (d) 2000

(2) Human Genome consists of 100,000 genes
(a) True (b) 10,000 (c) 30,000 (d) 50,000

(3) Match the following:

(a) Human Genome Project	(i) Baltimore
(b) Scotland	(ii) DNA
(c) Rosalind Franklin	(iii) NIH
(d) Reverse transcriptase	(iv) Dolly

(4) Match the following:

(a) Watson and Crick	(i) Semi-conservative replication
(b) Avery, MacLeod and McCarty	(ii) Base composition
(c) Meselson and Stahl	(iii) Double helix
(d) Chargaff	(iv) DNA is genetic material

(5) The physiological form of DNA is
(a) A DNA (b) B DNA (c) Z DNA (d) all the above

(6) A commercially standard substrate for lipase assay is
(a) Triacetin (b) Tallow (c) Olive oil (d) coconut oil

(7) Lipase consists of
(a) Ser-His-Asp (b) Ser-His-Glu (c) G-S-X-S-G (d) all the above

(8) Lipases are
(a) regiospecific (b) substrate specific (c) stereospecific (d) all the above

(9) Which of the following is a yeast lipase genera
(a) *Pseudomonas* (b) *Aspergillus* (c) *Candida* (d) *Rhizopus*

(10) Lipases are used in
(a) Fabrics (b) Food and dairy (c) both (d) none

Answers

1a 2c 3a-iii b-iv c-ii d-i 4a-iii b-iv c-i d-ii 5b
6c 7d 8d 9c 10c

Chapter 2

ISOLATION STRATEGIES OF MICROBIAL LIPASES

2.1. INTRODUCTION

Microbial lipases are generally the preferred sources for lipase production, and have been of research interests for the past several decades for basic research as well as commercial exploitation and still continue to be the choice of interest for biotechnologists pursuing in industrial and biochemical technology. Methods of screening for potent microbial sources have been considered an important step in any enzyme process with respect to its adaptation and habitat. Screening procedures are simpler to perform on a laboratory level and requires a sterile environment. Screening techniques involve a preliminary screening followed by confirmatory procedures and authentication of the enzyme productivities. Essentially, screening of microbial sources requires agar plate assays and submerged fermentation enzyme production at the flask-levels. Media selection is an important criterion for screening and identification that also contribute to the potency and induction of the enzyme activities in response to their habitual properties. Screening in agar plates are generally used for extracellular enzymes with enriched or inducible samples. The procuring of samples is another ideal step towards screening and may be from soil, natural waters, sludges or any byproduct materials.

Simple sampling and enrichment culture are sampling methods for the plate assays. The samples are screened on substrate/inducer agar plates by serial dilution and the screened isolates that are considered arbitrary producers are streaked to obtain pure isolates (aseptic technique). The arbitrary selection is based on the products of the reaction which may be coloured products or clearance zones obtained from the reaction. Various methods of agar plate assays are available in literature and could be standardized. Finally, the pure strain is tested for its ability to produce the enzyme and assessed for its titres of enzyme activities in comparison, generally, in submerged fermentation systems, supplemented with appropriate inducers or substrates.

We hereby discuss the different methods for screening of microbial sources for lipase production from bacteria, fungi and yeasts.

2.2. Sampling

2.2.1. Simple Sampling

This method of collection of samples requires field trips in farms or natural habitats where the desired microbial source might happen to be originated or adapted. Lipase sources generally reside in oil/lipid rich habitats, such as kitchen grease, fatty byproduct materials (leather fleshings) in slaughter house, oil refinery soils, etc. These samples can be directly used for the plate assays.

2.2.2. Enrichment Culture Technique

This method is generally an easier and a reliable procedure for screening of potent sources. The samples obtained from simple sampling are introduced into culture media in shake flasks with varying media compositions suited for bacteria, fungi and yeasts, with olive oil as an inducer. The log phase of the first cultures are inoculated into freshly prepared media (of the same composition) as starter cultures and consecutively inoculated, with increasing concentrations of olive oil (from 1% -5%). The final inoculation (say 4-5) culture is aseptically sampled for plate assays.

2.3. Screening Methods for Microbial Isolates

2.3.1. Lipase Plate Assays

Plate assays for lipase have been developed for easier monitoring and simpler methods of screening. Tributyrin has been universally employed for lipase screening. Tributyrin is a 4-C triglyceride with three butyric acid molecules in a glycerol backbone. Since lower fatty acids are sparingly water-soluble, this can be easily mixed with the screening media with slight sonication or homogenization. A stock of tributyrin is prepared in distilled water and added with appropriately prepared media. The samples are serially diluted on tributyrin agar plates and incubated at room temperature until halozones appear. The basic principle is that tributyrin substrate is hydrolysed by the lipase-producing microbial flora which produces a clearance zone in the agar plates visualized as a halo around the microbial isolate. This is due to the extracellular nature of most lipases and hence the activity can be correlated to its potency due to this feature. However, this is not a precise quantification but is qualitative. Olive oil substrates may also be used but with emulsification with PVA, generally. Other emulsifiers such as gum acacia, gum arabic, sodium deoxycholate etc. may be used but is declined in recent investigations. Rapid visualization however requires the use of Rhodamine B, which produces a halo and fluoresce in UV-irradiation.

Modifications of this procedure with use of $CaCl_2$ to precipitate with Tweens have also been successfully investigated. The fatty acids of the lipase reaction co-precipitate with Tweens to produce calcium oleates in wells of the microbial cultures. An indication of the precipitate similar to clearance zones is predictive of lipase activity/production.

2.3.2. Lipase Production in SmF

Subsequent to arbitrary selection from plate assays, the lipase producing microbial isolates are tested for confirmatory production in SmF at the flask-level in an orbital shaker under standard environmental conditions. In running a submerged fermentation, the most significant criterion is the media selection. The type of medium to be employed depends on the type of microorganism and the desired type of microorganism to be isolated. Generally, a standard medium could be used for culture such as semi-synthetic or minimal media for bacteria, or Czapek-Dox media for fungi. Essentially, the pH of the media maybe used to vary in order to obtain acidophiles, alkalophiles or in case of extremophiles, high temperatures have to be used during culture growth.

Another significant factorial design is the use of media suitable for specific microbial isolates whereby, previously designed media, for example, for *Pseudomonas* or *Rhizopus* may be employed. The various media composition are listed below which can definitely identify a suitable microbial strain, in case for a bacterial lipase producer:

***Medium A (M9 minimal medium) g/L**

Na_2HPO_4	6.0
KH_2PO_4	3.0
NH_4Cl	1.0
NaCl	0.5
#Glucose solution (20% w/v)	10.0 ml
$MgSO_4.7H_2O$ (24.65%)	1.0 ml
Thiamine.HCl solution (0.1%)	1.0 ml
$CaCl_2$ solution (1.47%)	1.0 ml
pH	7.0

For growth of bacterial sp.

***Medium B (Alkaline *Bacillus* medium) g/L**

Peptone	10.0
#Glucose	10.0
Yeast extract	5.0
K_2HPO_4	1.0
Na_2CO_3 (10%)	100.0 ml
pH	8.5

For cultivation and maintenance of alkalophilic Bacillus *sp.*

***Medium C (*Bacillus* broth) g/L**

$(NH_4)_2SO_4$	1.3
#Glucose	1.0
Yeast extract	1.0
KH_2PO_4	0.37
$MgSO_4.7H_2O$	0.25
$CaCl_2$. $2H_2O$	0.07
$FeCl_3$	0.02

pH	4.0

For cultivation of acidophilic Bacillus *sp.*

***Medium D (*Pseudomonas* basal minimal medium) g/L**

K_2HPO_4	12.5
KH_2PO_4	3.8
$(NH_4)_2SO_4$	1.0
$MgSO_4.7H_2O$	0.1
a#Carbon source (0.8 M)	100.0 ml
**Trace element solution	5.0 ml
pH	7.2

**Trace element solution	
Boric acid	0.232 g
$ZnSO_4.7H_2O$	0.174 g
$FeSO_4$ $(NH_4)_2SO_4.6H_2O$	0.116 g
$CoSO_4.7H_2O$	0.096 g
$(NH_4)_6Mo_7O_{24}.4H_2O$	0.022 g
$CuSO_4.5H_2O$	8.0 mg
$MnSO_4.4H_2O$	8.0 mg
a#Carbon source	14.4 g/100 ml

For cultivation and differentiation of Pseudomomas *sp. based on their ability to grow in different C-sources*

***Medium E (*Streptomyces* medium) g/L**

#Glucose	5.0
L-glutamic acid	4.0
KH_2PO_4	1.0
NaCl	1.0
$MgSO_4.7H_2O$	0.7
pH	7.0

For cultivation of Streptomyces kanamyceticus

* Media for lipase production should include olive oil as inducer at a concentration of 1.0% (v/v)

For lipase production, glucose should normally not be used as C-source, since lipases are susceptible to catabolite repression in the presence of glucose. Hence, either sucrose or a standardized C-source could be used.

There are umpteen numbers of other media for any given microorganism which can be referred with largely available literature. The media stated here have been abstracted from, Handbook of Microbiological Media, by Ronald M. Atlas, CRC Press (1993). In case of other microorganisms such as yeast or fungi, similarly, several other media exist, for example, a standard yeast medium is,

Yeast extract malt extract glucose medium (g/L)

Glucose	10.0
Neopeptone	5.0
Malt extract	3.0
Yeast extract	3.0

Likewise, glucose can be replaced by other C-sources and supplemented with olive oil as inducer.

There has to be a careful choice of medium and media constituents in the shake-flask cultivations. One or two components may be included or replaced as per desired productivities. For example, addition of trace element solutions is significant for bacterial sp., while for fungi, any complex medium would be ideal. An essential factor in the shake flask culture is the temperature which has to be suited for bacteria or fungi which are psychrophilic, mesophilic or thermophilic. For mesophiles, a temperature of 30°C or 37°C would be ideal under standard conditions. In case of psychrophiles or thermophiles, as their names imply, there should be a lower limit (4°–25°C) or an upper limit (40°–70°C). Thereby, the choice of a suitable microorganism depends on the media, culture conditions such as pH and temperature.

A number of other factors which have to be considerably studied are the rotation speed which is significant for aerobic cultivations. A static culture would produce little or no metabolite (lipase) under study for an aerobe and therefore shake cultures have to be employed, and the rpm of the orbital shaker have to be manipulated to produce optimal enzyme levels. Since most lipases are aerobic and produced extracellularly, supply of oxygen is correlated with rpm and also, under experimental conditions, absorbent cotton wools would be preferred to neck the flasks.

Another important parameter is the use of inducers in the media components. Olive oil is the most widely used inducer for lipase production in shake flasks at an optimal concentration of 1.0% (v/v). However, in case of requirements for further induction purposes, olive oil concentration may be increased, but this may also result in ambiguity since microbial isolates with poor lipase productivities may also be selected as a lipase producer. Apart from olive oil, a standard inducer, other commercially available oils such as coconut oil, castor oil, rice bran oil may be relevant as inducers.

The time dependence of the metabolite production (lipase) is a significant factor affecting 30–40% of lipase productivity. Generally, a continuous gradation of culture fermentation time could be employed in order to check the lipase production at intervals of 12 h or 24 h. The time factor depends on the basis of bacteria which may require shorter fermentation times (12–72 h), and in case of fungi and yeasts, may require prolonged periods even up to 144 h. This is an extremely important step for a successful isolation strategy. In other words, lipase production has to be checked at consecutive time points of 24 h, 48 h, 72 h, 96 h and so forth for a 24 h interval time period. Earlier time periods of 24 h and 48 h for optimal lipase production are suited for commercial exploitations.

Due to the extracellular nature of lipases, a simple filtration or centrifugation may be required for isolation of the crude lipase enzyme from the fermented broth. This is then followed by a lipase assay procedure usually a less time consuming assay protocol. Generally, spectrophotometric estimations would be desirable. For example, a standard p-NPP lipase assay would be ideal for most lipase producers, since p-NPP belongs to the higher chain fatty acid group (palmitate ester). The spectrophotometric detections can identify from 4-5 U of lipase activities. However, there is a possibility for misconceptions that esterases would be wrongly identified. A standard assay is titrimetry using a pH-stat, but identifications of a lipase producer would be in the order of more than 10 U. The assay protociols may vary for desired lipase properties, such as in case of dairy applications, β-naphthyl esters may be used as the substrate.

With all these ideas and suggestions for a successful and fruitful isolation strategy, it may still require that hands-on-experience would be vital to identify a suitable lipase producer with keen observations and suitability of samples that are utilized for the screening procedures.

2.3.3. Environmental Screening

Environmental screening is another important technique for isolation of microbes in the soil by analysis of their DNA. This is possible for screening of larger number of microorganisms present in the soil flora. This technique involves the construction of a soil DNA library wherein the desired metabolite producing microorganism could be obtained in contrast to the conventional screening methods in which the desired microorganism can be obtained. This procedure essentially is carried out from the isolation of soil DNA with CTAB. CTAB is used at varying concentrations to obtain pure DNA from soil. Further purification of the soil DNA is to be ensured by agarose gel electrophoresis. The soil metagenome is then formed into a library and similar to chromosome walking, the DNA of the desired metabolite is identified, also using probes of desired DNA sequence. This is a very good method of screening for a number of metabolites from the same DNA library which could be stored for several years. The identification of the microorganism is possible by DNA analysis of its phylogenetic origins. A large number of databases are available for identification of the phylogeny of microbes and hence, this method could be a potential tool for screening of microbial isolates. However, this method has been of little use in laboratories due to lack of awareness, and the difficulty in obtaining pure soil DNA samples. This is a common method for any metabolite, including lipases without requiring an inducible component and tedious fermentation processes.

2.4. Identification of the Microbial Isolate

Upon identification of a suitable microbial isolate for lipase production, it is deemed essential that identification of the microorganism with reference to its taxonomy is a significant criterion. The microbial lipase producer may be identified by references with Bergey's manual for classification by following suitable protocols. This is a tedious and expensive procedure and requires handling with most extreme care. With development and extensive laboratory practice, it has been made possible for commercial laboratories to identify microbial strains based on morphological and metabolic properties and also based on customer's specifications, wherever necessary, to successfully identify any given microbial sp. In this context for a lipase producing strain or any other strain, laboratories with culture collections undertake identification services on a payment basis. Commercial microbial identification has been on the rise and the internet offers a wide range of Laboratories suited for these services. Availing these services would be easier for customers to identify their microbial isolates. HiMedia has also produced several kits for identification of microorganisms which are commensurate with laboratory practitioners who want to identify their own strains of interest.

Apart from identification of a microbial isolate, its toxicity analysis should be followed prior to using the isolate for specific purposes (lipase production). For example, an *Aspergillus* sp. should be checked for its toxicity in case of the *flavus* species which produce aflatoxins. This species has been banned for study or any investigation in some countries like Japan. A simple method for identification of the *flavus* variety has been referenced by Atlas (1993):

Aspergillus differential medium (%)

Pancreatic digest of casein	1.5
Yeast extract	1.0
Ferric citrate	0.05
Agar	1.5

For the cultivation and differentiation of Aspergillus flavus, *which appears as bright orange colonies.*

Thereby, a successful isolation strategy involves a careful series of steps and has to be considered an important step for commercial exploitations of microbes and their metabolites.

Sampling (direct/enrichment)

↓

Serial dilution

↓

Plate assays

- Spread plating (with tributyrin)
- Streak cultures (with tributyrin)
- Aspetic cultures (to obtain pure isolate)
- Rhodamine B/Victoria blue plate assay

↓

Maintenance of cultures (on LB/CDA agar)

↓

SmF (in shake flasks with inducer)

↓

Identification of microbial isolate

↓

Toxicity testing

Figure 2.1. An isolation strategy for a lipase producing microbial isolate.

Self-evaluation

(1) A preliminary screening for microbial isolates is
(a) spectrophotometry (b) plate assays (c) Radiolabelling (d) HPLC
(2) Screening procedures by plate assays could be used for
(a) Lipases (b) Proteases (c) Cellulases (d) all the above
(3) A plate assay for lipase can use the substrate
(a) Tributyrin (b) Olive oil emulsion (c) $CaCl_2$-Tween 80 (d) all the above
(4) Submerged fermentation for lipase production should include
(a) Tributyrin (b) Olive oil (c) p-NPL (d) Glucose
(5) Environmental screening is based on
(a) DNA (b) Temperature (c) pH (d) Inducer

Answers

1b 2d 3d 4b 5a

Chapter 3

LIPASE PRODUCTION: DEFINING THE MEDIA

3.1. INTRODUCTION

For any commercial production of a metabolite or an enzyme from a microbial source of origin, an important process is the fermentation process whereby, microbial cultures grow in a medium, an environment to suit its metabolic growth for the desired end product. Generally fermentation processes are carried out in shake flasks in lab-levels and up-scaled in large vessels, known as bioreactors. This process of conversion from a single cell or an inoculum of cells to produce enzymes, in the presence of varied amounts of nutrients under favourable environmental conditions is called, fermentation.

Technically, 'fermentation' is an industrial process for the breakdown of complex substrates into metabolites in a culture vessel with sufficient nutritional requirements and oxygen supplementation requisite for proper growth of the microorganism (in terms of biomass) to yield its product, often an independent variable with increasing biomass. In the case of lipases, this process is of considerable importance to enable the microbial source to grow in a medium supplemented with inducer (olive oil) for a period of 24–72 h or longer to yield lipase. In other words, the media components are utilized for growth of the microbe and in the presence of the inducer, an upregulation of lipase production takes place and this enzyme is released into the culture medium extracellularly. Now, this process of production is inducible and the lipase is produced in limited amounts. A process known as optimization is required to enhance the optimal production of lipase by optimizing the amounts of variables which are the factors responsible to increase the amounts of lipase productivities. These variables may be N- and C- sources, salts, inducers, and environmental variables, pH and temperature. Thereby, the optimization of a medium into an economically viable process with optimal formulations to suit commercial production is a vital process in any enzyme production study. We will describe this process of medium formulation in this chapter with brief notes on the types of media, metabolic factors and various modes of culture growth responsible for the overall fermentation process.

3.2. Fermentation Media

The fermentation medium is one that determines the chemical or nutritional environment essential for microbial growth and hence, the end product. The medium employed for a microbial fermentation depends on the type of the microbial source, be it a bacteria, fungi or yeast. It is therefore a critical component of any industrial fermentation process, directly affecting not only the productivity but also process economics.

A 'medium' in any fermentation process could be defined as a chemical or complex nutrient system suited for exponential microbial growth so as to increase its metabolic rate for sufficient product synthesis. In this context, depending on the medium constituents, fermentation is broadly classified into submerged fermentation (SmF), wherein the nutrients are suspended or dissolved in a liquid phase, usually water, and solid state fermentation medium (SSF) wherein, the medium is a solid phase, which may be of agricultural origin or any by-product, such as molasses. In this chapter, we have attempted to describe lipase production in SmF system which is common to typical microbial cultures, while SSF is favoured for the fungal cultures.

Based on media constituents, the fermentation medium may be complex, synthetic or semi-synthetic formulations. Complex media are generally natural extracts containing mixtures of components of varying compositions. Yeast extract is a rich source of vitamins and has widely been used in laboratory as well as commercial fermentation processes. However, it is undefined in composition and may also vary from each manufacturer. Peptones are similarly complex and may be neo-peptones or bacto-peptones depending on the type of extracts from and the manufacturer. Peptones are usually tryptic-digested extracts from animal or milk origin for use in microbiological media. Another form of peptone is tryptone which is a Type I Casein peptone. Meat or beef extracts are also complex media containing high proportions of nitrogenous constituents. Complex media have the ability for the growth of most microbial cultures and can also be used for the growth of unidentified microbial sources.

In contrast, synthetic media are those whose chemical composition is defined and the amounts of chemical constituents employed correlate with the stoichiometry of metabolic growth rates of the microbial cultures. Therefore, synthetic media, wherein no natural extracts are included defines the composition of the medium and any addition or deletion of any constituent can be well-analysed to its effects on the microbial growth. The synthetic media or chemically defined media have high prospects in fermentation studies and are preferred in laboratory experiments for analyses of variables that determine the fermentation process. Chemically defined media contain nitrogen sources such as ammonium salts, carbon sources such as glucose (to be avoided for lipase), phosphates for buffering, metallic salts as trace elements and if necessary for some fermentations, vitamins. This defined composition allows one to monitor the fermentation process and determine the exact amounts added and analyse the stoichiometric depletion of these sources during a fermentation course.

Semi-synthetic medium is used as a metaphor to compare and combine the features of complex and chemical media to achieve better fermentation processes for higher product yields. As the name implies, along with a chemical medium, a complex substrate is introduced to provide additional nutrient supplements to enable rapid growth of the microbial culture and improve the product yield. This has reference to most microbial fermentations used currently with and is of practice in most laboratories for fermentation studies. Therefore, the semi-synthetic medium may contain complex nitrogen sources such as peptone, complex carbon sources such as vegetable oils along with chemically defined media. Yeast extract is an excellent supplement of vitamins to the chemical medium.

Another significantly used medium is the use of minimal media. A minimal medium is one that contains the barely essential requirements for a microorganism to grow on defined amounts of media constituents. Usually minimal media are more suited for bacterial fermentations, due to their speed of growth in a medium with a basic amount of N- and C-sources along with essential salts just enough for the growth at the metabolic rate with concomitant depletion of the media nutrients as the fermentation period progresses. However, this should not be usually considered for the course of investigations to improve product yields. Physiological factors require adequate supply of nutrients and the minimal media may safely be disregarded. Minimal media defines the medium constituents and is therefore a potential media for bacterial fermentations requiring stoichiometric studies of growth rates. The different types of media constituents forming the various media are indicated in Table 3.1.

Table 3.1. Types of Media Formulation

Complex	Defined	Minimal	Semi-synthetic
Beef extract	NH_4Cl	sucrose	Casamino acids
Yeast extract	KH_2PO_4	K_2HPO_4	Yeast extract
Peptone	$MgSO_4\ 7H_2O$	KH_2PO_4	Trisodium citrate
Glucose	$CaCl_2\ 2H_2O$	$(NH_4)_2HPO_4$	KCl
	Elemental Sulfur	$MgSO_4\ 7H_2O$	$MgSO_4\ 7\ H_2O$
	CO_2	$FeSO_4\ 7H_2O$	$FeCl_2$
		$MnSO_4\ 7H_2O$	NaCl

3.3. Media Optimization with Nutritional Factors

The choice of media is unique for different microorganisms and varied combinations of media components are required in order to use the microorganisms for full benefit. Hence, a procedure is necessitated to optimize the various factors to be used as medium constituents. The medium in which the microbe attains good cultural requisites can be preferably optimized. For example, a synthetic medium for bacteria in which it produces the lipase of about 5-10 or more units can be optimized. Each constituent of the synthetic medium needs to be optimized and its response is observed. Consider the optimization of an N-source, urea used in the previous medium. Urea has to be used in the control medium while, other inorganic N-sources such as ammonium chloride, ammonium nitrate, ammonium citrate, ammonium hydrogen phosphates, ferrous ammonium sulfate can be used in test medium at the same concentration as urea and the N-source yielding an increase in lipase activity can be further optimized at its various concentrations. Similarly, there are a number of variables (C-sources, inducers, salts) which can be optimized for producing optimal or the maximum amount of lipase. These factors are the nutritional requirements for the microbial source. The C-sources may be sugars such as maltose, lipids such as triglyceride esters or even vegetable oils. However, the appropriate C-sources are the carbohydrates-hexoses, pentoses, polysaccharides like starch, cellulose. In the case of lipases, lipids or oils can be used as inducers which are responsible for their inducible production. A medium without any inducer will greatly hamper the lipase production. This is because, most lipases are inducible. Depletion of N- and C-sources will also not result in increased lipase productivity since these are the most essential components for metabolic turnover of the microbial cell. This means that these sources are required for cellular metabolism of the lipase producer or in any case, any microbial source. N-sources aid in biosynthesis of proteins and nucleic acids. C-sources are mostly responsible in the respiratory transport by way of glycolysis and yield of ATP molecules is used in oxidative phosphorylation. Salts are as well important additives in a medium especially for buffering action by phosphates (potassium phosphates are ideal). Salts such as Mg, Ca, Na are essentially the macronutrients, while Mn, Zn, Mo, Fe are the micronutrients grouped as trace elements. A trace element stock is prepared and an aliquot is added to the larger volume of the medium due to the trace quantities required by the microbial cell. These macro- and micro-nutrients are important components of a medium, however, Ca, Fe may be inhibitory in certain cases. In case of optimization of a synthetic medium, supplementation with vitamins are essential for a bacterial culture. However, vitamins should not be sterilized in an autoclave but added prior to inoculation by filter-sterilization. These vitamins are required as coenzymes and cofactors for metabolism. Vitamins can also be provided by addition of yeast extracts, in this case, it is a semi-synthetic medium because the vitamin formulations are not defined. Therefore, the nutritional requirements for a microbial culture is absolutely essential not only for the cellular growth but also for optimal production of the lipase enzyme, or generally any other metabolite (Table 3.2).

Table 3.2. Functions of Nutrient Factors in Microbial Cellular Metabolism

Nutrients	Function
Carbon	Main constituent of cellular material
Oxygen	Constituent of cell material and cell water; O_2 is electron acceptor in aerobic respiration
Nitrogen	Constituent of amino acids, nucleic acids nucleotides, and coenzymes
Hydrogen	Main constituent of organic compounds and cell water
Phosphorus	Constituent of nucleic acids, nucleotides, phospholipids, LPS, teichoic acids
Sulfur	Constituent of cysteine, methionine, glutathione, several coenzymes
Potassium	Main cellular inorganic cation and cofactor for certain enzymes
Magnesium	Inorganic cellular cation, cofactor for certain enzymatic reactions
Calcium	Inorganic cellular cation, cofactor for certain enzymes and a component of endospores
Iron	Component of cytochromes and certain nonheme iron-proteins and a cofactor for some enzymatic reactions

However, cellular growth and enzyme production are not linear with increasing time periods and therefore optimal production is not exponential. Inspite of this fact, in many cases, metabolite production also depends on cellular growth due to which primary metabolites and secondary metabolites are classified according to the growth phase of the microbial culture. This is in itself a comprehensive analysis and is not mentioned here which is beyond our scope. Nutritional requirements are vital to any microbial culture which has to be exploited to maximize metabolite production. Nutritional optimization does not fulfil all the criteria for lipase production generally in a culture vessel. There has to be oxygen supplementation or in terms called aeration which is also a deciding factor for cell growth and is a determinant of facultative/obligate aerobes or anaerobes. Generally most of the microorganisms used for commercial purposes are aerobic. In shake flasks in laboratories, aeration is provided by orbital shakers and the speed of the rotations can be monitored and in certain cases, be optimized. Oxygen can also be considered a nutrient factor which forms part of the optimization process.

The optimization studies of nutrient requirements can be conventionally carried out in shake flask experiments with one variable-at-a-time, and this is highly time-consuming and also tedious. Recent methods utilize factorial designs wherein more variables can be studied in a single and a series of experiments. This means that simultaneous use of C-source, N-source, inducers and any other variable can be optimized in a pre-calculated set of experiments. This is highly advantageous in terms of time and work. This may increase the lipase productivity to a desired limit. Hence, it requires the investigator to decide upon a factorial design by arbitrary choice of factors or a conventional design and in any case eventually yield higher lipase productivity.

3.4. Modes of Fermentation Process

3.4.1. Batch Culture

Batch culture is the conventional laboratory shake flask culture where the nutrients are exhausted during progression of growth. This means that utilization of medium nutrients is available only for a period of time during the growth phase of the microbial culture and until the lag phase. The nutrients are assimilated and metabolized for the culture growth of the microorganism while it produces its metabolites at specific growth rates. This is achieved by preparation of the medium, sterilization, inoculation, incubation followed by extraction of the metabolite. No further nutrients are supplied at any phase of microbial growth. Once the medium is depleted of its nutrients, the separation of the metabolites takes place at time intervals, and, the media becomes unusable for further culture and has to be discarded. Thereby, batch culture, as the name applies is performed in batches of laboratory flasks with 50-100 ml media. The first batch is used up and the second batch follows. This is generally the way of procedures for optimization of media components and is a widely used fermentation process. In the batch culture all the organisms are present at different stages of growth or the cells grow randomly. Such bacterial populations do not give conclusive studies about the growth behavior of individual cells, as the distribution of cell size and cell age is random. Hence, other modes of culture growth have to be followed for growth behavior.

3.4.2. Synchronous Culture

In contrast to batch culture, studies involving growth behavior require that the microbial cells are in the same phase of the growth curve, that is exponential phase or the log phase. This method of growing a microbial cell culture of the same phase is called as the synchronous culture. All the cells in the synchronous culture will divide at the same time, all will grow for the same generation time, and all will divide again at the same time. Thus the entire cell population is uniform with respect to growth and division. Inorder to obtain synchronous cultures, different techniques are available which include:

- Manipulation of environmental conditions (Induction technique),
- Separation by physical means of cells in population that are at the same stage of the cell cycle (Selection technique) and,
- Helmstetter—Cummings technique.

It is not always possible to analyze a single bacterium because of its small size. It is very important to know about cell organization, differentiation and macromolecule synthesis. Results made for synchronized cultures are therefore equivalent to the results made for individual cells. However, the synchrony lasts only for a few generations, since even the daughter cells of a single mother cell soon get out of phase with one another. Synchronous cultures are therefore not used in day-to-day laboratory practice except for studies on growth behavior.

3.4.3. Fed-batch Culture

Fed-batch fermentation is a production technique in between batch and continuous fermentation. A proper feed rate, with the right component constitution is required during the process. Fed-batch offers many advantages over batch and continuous cultures. From the concept of its implementation it can be easily concluded that under controllable conditions and with the required knowledge of the microorganism involved in the fermentation, the feed of the required components for growth and/or other substrates required for the production of the product can never be depleted and the nutritional environment can be maintained approximately constant during the course of the batch. The feed is usually the rate-limiting factor in a fed-batch. In case of lipases, inducers can be the rate limiting factor for an increased productivity. Sometimes, controlling the substrate is also important due to catabolic repression. Hence, glucose should not be the limiting nutrient for a lipase. Fed-batch usually permits the extension of the operating time and high cell concentrations can be achieved, resulting in an improved productivity [mass of product/(volume x time)]. This aspect is greatly favored in the production of growth-associated products. The controlled periodic shifts in growth rate provide an opportunity to optimize product synthesis, particularly if the product of interest is a secondary metabolite whose maximum production takes place during the deceleration in growth.

3.4.4. Continuous Culture

This system reflects that a microbial population can be maintained in the exponential growth phase and at a constant biomass concentration for extended periods. This is called a steady-state or balanced growth. Continuous culture is common in nature, for example, the GI tract houses the microbial flora which is continuously replenished with nutrients. Chemostat is a simple device to maintain a continuous culture. It consists of a vessel in which the culture is grown. Devices are connected to the growth vessel to control the amount of air and fresh medium to be added to the culture as well as the overflow of medium containing cells and waste products (Figure 3.1).

The culture reaches to a constant volume and density when equilibrium between inflow of fresh medium and outflow of medium reaches. All the constituents of medium, except one, are in excess in chemostat. The deficient nutrient will support the growth of only limited number of micro-organisms. Thus the density of culture in chemostat is controlled by nutrient limitation. Sucrose as a carbon source can be used in setting up chemostat. By controlling the rate at which nutrients are added to the chemostat, the rate of bacterial growth is also controlled. The rate of addition of nutrients is known as dilution rate, if dilution rate increases, more bacteria are washed out of culture vessel than are produced by cell division.

Continuous culture technique provides a constant source of active cells in the exponential phase which are highly useful in fermentation industry because of their high physiological activities. As the cells are grown continuously, it is possible to study synthesis or catabolism of limiting substrates and also to select various classes of mutants. Production of single cell protein (SCP) can also be achieved by this culture method.

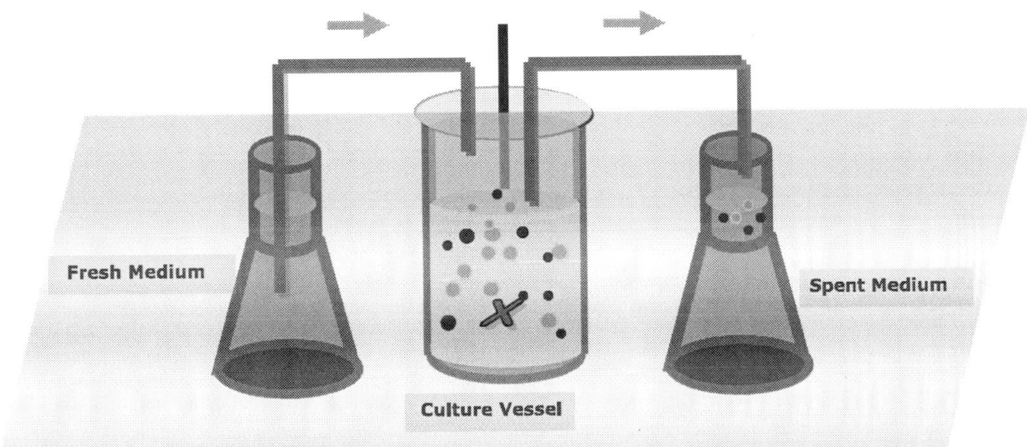

Figure 3.1.A basic chemostat.

3.5. GOOD MANUFACTURING PRACTICE (GMP)

GMPs are a set of legal codification of sound quality principles, which have been used by the pharmaceutical and healthcare manufacturing industries for over 50 years as a means of assuring that products have the identity, strength, purity and quality that they claim to contain. GMPs are in effect in over 100 countries, and GMP compliance is a pre-requisite to exporting bioprocessed products between countries. GMP takes the holistic approach of regulating the manufacturing and laboratory testing environment itself. An extremely important part of GMP is documentation of every aspect of the process, activities, and operations involved with drug and medical device manufacture. If the documentation showing how the product was made and tested is not correct and in order, then the product does not meet the required specification and is considered contaminated and unsafe for use. Additionally, GMP requires that all manufacturing and testing equipment has been qualified as suitable for use, and all operational methodologies and procedures. GMP compliance is an essential part of every bioprocess facility and the production criteria should pertain in accordance with the set of manufacturing practices.

Self evaluation

(1) A complex medium constituent is
(a) Beef extract (b) Yeast extract (c) Tryptone (d) all the above
(2) Addition of 100% water is essential for
(a) SSF (b) SmF (c) both (d) none
(3) Lipase fermentation should not include
(a) Lactose (b) Glucose (c) Olive oil (d) Coconut oil
(4) Optimization of medium is
(a) to increase production(b) to decrease activity (c) to increase pH
(d) to increase biomass
(5) Addition of rate-limiting nutrient at time intervals to the medium is
(a) batch (b) synchronous (c) fed-batch (d) continuous

Answers

1d 2b 3b 4a 5c

Chapter 4

LIPASE PURIFICATION AND CHARACTERIZATION

4.1. INTRODUCTION

All fermentation processes yield the desired enzymes in a crude form usually not usable for structure-activity studies. The requirement of a protein in its purest form is an absolute criterion for most studies. The crude enzyme can be used for commercial preparations in terms of industrial applications, and purification procedures may be an additional burden regarding cost and yield of enzymes. However, when one has to investigate structure analyses and structure-function relationships, the production of a purified form of the enzyme is the first step towards these studies. The purified enzyme has to be crystallized for structure studies or its crystal properties. In terms of proceeding with a purification process, one has to bear in mind some of the essential requirements and feasibility regarding the type of purification procedures that are attributed to its enzymatic characteristics.

An ideal purification procedure would be successful when the pure enzyme has increased specific activity when compared to its crude form. This is the most significant feature of any purification method involved which will ultimately determine an efficient purification procedure.

Box 4.1. Specific Activity of a Protein

Specific activity of a protein is defined as the amount of pure protein of interest found in a large population of other proteins. Specific activity of a protein determines the purity of a protein and is calculated based upon its activity and the total protein content.

$$\text{Specific activity (per mg protein)} = \frac{\text{Activity of protein (U/ml)}}{\text{Total protein (mg/ml)}} \text{ Units /mg}$$

The specific activity of a protein can be visualized as a number of marbles present in a system and with increasing purity of the protein, the number of other proteins are eliminated. The specific activity of the protein of interest increases with reference to its higher fraction in the total system, but having the same activity.

Another aspect of a purification step is the purification fold of the pure enzyme. The derivations of the parameters in a purification step would be discussed in section 4.6 of this chapter.

One of the most deciding step for the choice of a purification protocol is the preservation of its native form and hence its activity. Preservation of the native form that is its native conformation is the fundamental aspect for any purification protocol and thereby an estimate of its activity is essential for enzymology practices. For example, when you want to study the characteristics of a pure lipase, such as positional specificity, an integral feature of most lipases, retaining the activity is very much essential. However, in a purification step, it is important to notice that upon employing subsequent steps of purification, the activity is lowered in each step. That is the total activity for the particular protein of interest (lipase), gets reduced as the purifications steps increase. Therefore, a minimization of the number of purification steps would be advisable. Recently, one-step purification protocols are widely used in most laboratories for efficiency, cost and speed. This is however, not a stringent criterion for all proteins. The loss of activity at each step of purification, also implies that there is some loss of the enzyme at each step. This may be due to presence of proteases that degrade the proteinaceous structure, detergents that destabilize enzyme conformation, rise in temperature that denatures enzyme activity, or careless handling of the enzyme during dialysis and centrifugations.

Time constraint is an important factor that is the ultimate aim of most laboratories. Speed of purification is a determinant of competency amongst most researchers, but it is also necessary that one has to be first familiarized with the nature and properties of the protein that is evident from literature. For example, it would be wise to use an affinity procedure for glycoproteins (eukaryotic lipases) to exploit the specificity of binding with Con A matrices to mannose residues of the enzyme, or a hydrophobic column that exploits its hydrophobicity, rather than follow an ion-exchange protocol that is to be employed for ionic proteins. Under rare conditions, this may suit the purification process and with recent developments in the availability of combined ion-exchange and hydrophobic matrices like Q- Sepharose. For larger proteins, an initial gel chromatography would be ideal even without requiring concentration of the enzyme with ammonium sulfate, since large proteins would be eluted first in a gel matrix and is a purified as well as a concentrated protein solution. Keeping such points in mind and with extensive analysis of literature, it would finally speed up the purification process for a successful strategy of obtaining a pure protein.

Cost of purification is of little significance in purification of proteins for structure studies and therefore, high cost matrices can be used. Most matrices are supplied in a ready column packaging and cost is generally not a deciding factor. Consider buying a resin that has to be activated with acid or alkali overnight or swelling up of a gel resin overnight, or activating a ligand to an affinity matrix, followed by washing procedures, and if carried out improperly would hamper the entire purification procedure. This could conveniently be replaced by pre-swollen matrices like DEAE-Sepharose or Sepharose CL-4B or or Con A-Sepharose. The use of Sephadex (Dextran) matrices could be safely avoided unless for concentration procedures, and can be replaced by Sepharose (Agarose) matrices for convenience, and not consider the cost factor. However, in large-scale protein preparations for commercial applications, cost of purification is a fundamental factor affecting the whole part of the application costs. Hence, there is a need for use of other methods such as partial purification using ammonium sulfate precipitation and simple filtration techniques with a concentrator. Aqueous two-phase partitioning system (ATPS) is a very good method of choice for commercial purifications which could be PEG/salt or detergent partitioned. Herein, the protein of interest partitions itself into the PEG phase or salt phase depending upon its nature where, it could be separated with ease of extraction by simple centrifugation. In a detergent ATPS, the protein partitions itself into the detergent-rich or detergent-depleted phase, with rise in temperature of the cloud point of the detergent employed. Triton X-114 is an excellent detergent system for large-scale lipase purifications, and prokaryotic lipases partition to the detergent phase while eukaryotic lipases partition into the aqueous phase due to complexity of glycosylations.

Table 4.1. Protease Inhibitors

Protease inhibitor class	Inhibitor	Effective concentration
Serine inhibitors	Diisopropyl fluorophosphate (DIPF)	0.1 mM
	Phenyl methyl sulphonyl fluoride (PMSF)	0.1– 1.0 mM
Cysteine inhibitors	Iodoacetate (IAA)	10 – 50 μM
Aspartic inhibitors	Pepstatin A	1 μM
Metallo inhibitors	Ethylene diamine tetra acetate (EDTA)	1 mM

4.2. Preserving Enzyme Activity

Preservation of enzyme activity is the utmost ambition of any purification strategy. When dealing with enzymes that are intracellular or membrane-bound, upon isolation procedures of the initial step the enzyme is exposed to harsh conditions that will eventually lead to denaturation if left uncared. Also, for extracellular enzymes, the enzymatic nature of the protein has to be preserved throughout the purification process. Hence, a successful purification should undertake the preservation of enzymatic activity or the native form of the protein. Extreme pH and temperature are the principal factors causing denaturation. It is always better to perform the purification in a cold room at 4°C and at functional pH of the enzyme. The protein is generally suspended in a buffering medium to retain the pH of the enzyme. Tris.HCl buffers are very much ideal for purification processes, if the enzyme is active with this buffer solution. A screening may be performed initially to choose the buffer system to avoid ambiguity from literature, for which the enzyme under study may be inactive. Sometimes near neutral pH (7.0–8.0) of the buffers often work out well. Regarding temperature, it is universal that cold conditions have to be maintained right from concentration of the enzyme to the analysis by PAGE. An exception for lipase is that RT would be ideal only for performing the hydrophobic chromatography with octyl or phenyl Sepharose. However elution of the lipase with a fraction collector has to be carried out at 4°C. In addition to pH and temperature, proteases are very much troublesome hindering the purification process by inactivation of the enzyme. Therefore protease inhibitors have to be used in the protein solution. It is an important consideration, that lipases are Ser-enzymes similar to proteases having an active Ser residue at its active site. Hence, cleaving the Ser-protease would also cleave the Ser-lipase. Hence, it is even better to rule out the use of serine inhibitors or use metallic protease inhibitors, since, lipases usually do not require metals as cofactors. Use of protease inhibitors can be in buffer solutions in which the protein is dissolved. Table 4.1 lists out the different protease inhibitors for use in protein purification.

4.3. PROTEIN CONTENT

Protein content of an enzyme is the relative concentration of any protein present in a mixture of protein solution. An estimate of the total protein content is a measure of the specific activity of the protein or enzyme of interest. Thereby, protein estimation is an indirect method to analyse the amount of a particular protein in a given sample. There are several methods to determine protein content, each varying in sensitivity and accuracy.

4.3.1. Spectrophotometry

This is one of the oldest yet widely used method for estimation of proteins in the order of 50-1000 μg. This method utilizes the absorption of a protein solution at 280 nm due to presence of tyrosine and tryptophan residues. An aliquot of protein solution in a cuvette is measured at A_{280} in a UV spectrophotometer and the absorbance values are observed. Using a standard curve of known protein concentrations (20-200 mg BSA), one can determine the amount of protein present in the unknown solution. This method relies on the fact that all proteins absorb at 280 nm. However, this is not an accurate method to analyse the protein content during the purification steps. All the more, this method is very much appropriate to observe that all the protein has been eluted from a column after elution when the A_{280} becomes nil. Ammonium sulfate or buffers do not interfere with this method. However, this is a common procedure in most laboratories to save time when doing a column run to observe only concentration changes. A graphical plot can also be obtained along with the activity profile curve.

4.3.2. Lowry's Method

This is widely used for clinical enzymes such as Acid or Alkaline phosphatases, SOD, catalase, GST and is highly a sensitive procedure which can detect proteins in the range of 0.1-1.0 mg/ml concentration. The method uses an alkaline copper tartarate solution and with addition of Folin-Ciocalteu reagent produces a blue-colored complex, a reduction product heteropolymolybedenum blue from phosphomolydotungstate, which is measured at 640 nm. The intensity of blue colour is measured by absorbance at 550 nm for higher protein concentrations (100-2000 μg/ml) or at 750 nm for lower protein concentrations (below 500 μg/ml). Reducing agents and reducing sugars interfere with the Lowry assay. However, during purification procedures, this is a laborious and time-consuming experiment and needs to be replaced by other procedures, such as the Bradford assay which can be carried out in a matter of few minutes. Also, a microassay of the Lowry version has been used in certain laboratories.

4.3.3. Bradford's Dye-binding Method

This is a simple and efficient method for estimation of the protein content with speed, accuracy and sensitivity. The Bradford's procedure is a simple addition of Coomassie Brilliant Blue G-250 to the protein solution which forms a blue-colored complex that shifts the λ of the dye from 465 nm to 595 nm in the presence of *o*-phosphoric acid. The blue complex is stable for 1h and is a direct measure of the protein content. This method is sensitive to 1 μg of protein and can be used for a wide range of proteins without interference of non-protein components. However, the detergents Triton X-100 and SDS hinder this estimation protocol. It is also worthwhile to note that glass cuvettes are ideal for use of measurement of proteins and disposable glass cuvettes are advantageous. However, quartz cuvettes which may also be used need cleaning with HCl or methanol after use. This is a relatively rapid method in preference to Lowry's protocol. Improvements of Bradford's method have been made by addition of phenol to increase sensitivity and the use of a microassay that can detect up to 1-20 μg.

4.3.4. Bicinchoninic Acid (BCA) Method

BCA method is of a little commonly used protocol, however with high sensitivity. Inspite of its slow monitoring of protein content, it is similar to Lowry estimation in formation of a Cu^{2+} -BCA complex which can be read at 562 nm. The apple green color of BCA forms a purple colored complex of copper and BCA reagent and is a measure of the protein content. This is advantageous for proteins solutions that contain 1% detergents. The range of sensitivity of the BCA method is of the order of 1 μg.

4.4. Concentration of Protein

4.4.1. Ultrafiltration

Ultrafiltration is a method for concentration of target protein from the other protein components, based upon using a membrane filter. The separation is based on the size, shape and /or charge of the molecular species. Membrane filtrations can be used for desalting buffers and clarifications of turbid solutions. The membrane has to be chosen on the type of cut-off required from a previous knowledge of the protein MW. Ultrafiltration is a rapid and efficient method for concentration of large volumes of protein solutions obtained from a fermentation broth. It is a continuous process of filtration with flow rate adjustments. Membranes recently used for protein solutions are PVDF, cellulose acetate and cellulose nitrate, MW cut-offs ranging from 100-1,000,000 kDa and with controlled pore sizes of 0.025-15 μm. Ultrafiltration using these membranes for protein concentration requires suction pressure or centrifugal force for liquid flow. Use of suction pumps is common in most laboratories.

4.4.2. Addition of a Dry Matrix Polymer

This is a reasonably simple and quick method for concentration of protein solutions. A dry inert matrix of Sephadex G-25 is allowed to swell in the protein solution overnight at 4°C and the pores of the gel matrix are small enough to allow the target protein to pass through. The resulting protein solution containing the desired protein component is then separated from the protein mixture that is absorbed by the matrix upon gravity settling, filtration or centrifugation. This is an old method of concentrating a protein but has its advantages in terms of removing small cell debris and particulate solids.

Box 4.2. Concentration of Protein Using Dry Sephadex G-25

(1) 20 g Sephadex G-25 is added to 100 ml of the protein solution and stirred well. The gel matrix is allowed to swell overnight at 4°C. The resulting swollen mix is subjected to gravity settling and the remaining concentrated protein solution is recovered.
(2) Small volumes of 10–500 μl protein solutions can be concentrated in a microfuge with known amount of dry Sephadex G-25. Protein solutions can be concentrated upto 5-fold upon centrifugation and collected through a pinhole at the base of the microfuge. Protein recovery is around 90–95 %.

4.4.3. Ammonium Sulfate Precipitation

Concentration by ammonium sulfate precipitation is the most commonly used protocol in most laboratories due to ease of performance and efficient protein recovery in native conditions. This method is basically a salting-out method based on the presence and interactions of hydrophobic residues of the protein with ammonium sulfate. Water that is forced into contact with hydrophobic residues in the interior of the protein, becomes ordered. Upon addition of ammonium sulfate, water solvates the salt ions and with increasing concentrations of the salt, water is removed from around the protein, resulting in an exposure of the hydrophobic patch. The exposed hydrophobic patches on one protein molecule interact with those of the other molecules resulting in aggregation of many other protein molecules. This aggregation of proteins with varying numbers of hydrophobic patches result in fractionation of proteins, a principle for the concentration of proteins from other protein components. Addition of ammonium sulfate in increasing proportions has to be carried out at 4°C and it is of criteria that it has to be added in its salt form without prior dissolution under rapid stirring in a magnetic stirrer preferably. A Table for the addition of grams of ammonium sulfate to the protein solution in terms of increasing concentrations has been determined (refer Section II, Appendix II). This Table gives reference as to how much ammonium sulfate is needed for increasing its concentration from 0% to 100%. For example, an initial addition of the salt would be from 0% to 25%. With reference to the Table, the required salt is added appropriately and left overnight for precipitation conventionally. The precipitate is collected by centrifugation at 8000-10000 rpm in a cold centrifuge at 4°C and the supernatant is concentrated to 50% ammonium sulfate saturation. The steps are repeated to 75% saturation and 90% saturation of the resulting supernatants. The precipitates are tested for the protein or enzyme activity and subsequently used for running a column chromatography. Prior to performing a column chromatography such as ion-exchange, the salt has to be removed completely, unless otherwise for certain procedures such HIC, which is an important purification protocol in case of microbial lipase enzyme purification. The removal of salt (ammonium sulfate) is by a simple dialysis against a suitable buffer repeated twice. The salt ions are removed from the protein and replaced by buffer ions used for the column run.

4.4.4. Precipitation by Organic Solvents

Precipitation by addition of organic solvents is a conventional method of concentrating a protein. Addition of the organic solvent lowers the dielectric constant of the solution, and hence its solvating power. The solubility of the protein is decreased and aggregation of protein molecules through electrostatic attractions can occur. This is the fundamental concept of precipitation using organic solvents. It must be considered that use of organic solvent precipitation should be carried out at 0°C, or loss of protein conformation at higher temperatures would be effected. Some organic solvents may result in harmful denaturation of the target protein. However, certain proteins are not susceptible to denaturation by water-miscible solvents such as acetone, ethanol and methanol and propanol. In such cases, proteins can conveniently be concentrated by using organic solvents. An activity analysis of the crude protein sample in a number of solvents can give an idea about which solvent can be used.

4.4.5. Precipitation with PEG

The precipitation of proteins using a high MW polymer, PEG, is another process of protein concentration depending upon the MW of the polymer to be used and can be removed by ultrafiltration or dialysis with suitable solvents to remove and replace the polymer. An advantage of using PEG is that it does not interfere with subsequent purification steps.

4.5. Purification by Column Chromatography

4.5.1. Ion-exchange Chromatography

Ion-exchange chromatography basically involves the separation of an ionic protein mixture by adsorption on to an ionic matrix which has opposite charges to the protein of interest and during elution, allows the sample to be eluted by using a stronger ionic solvent that is attracted preferably by the resin. In principle, the ion-exchange process of adsorption and elution is represented in Figure 4.1.

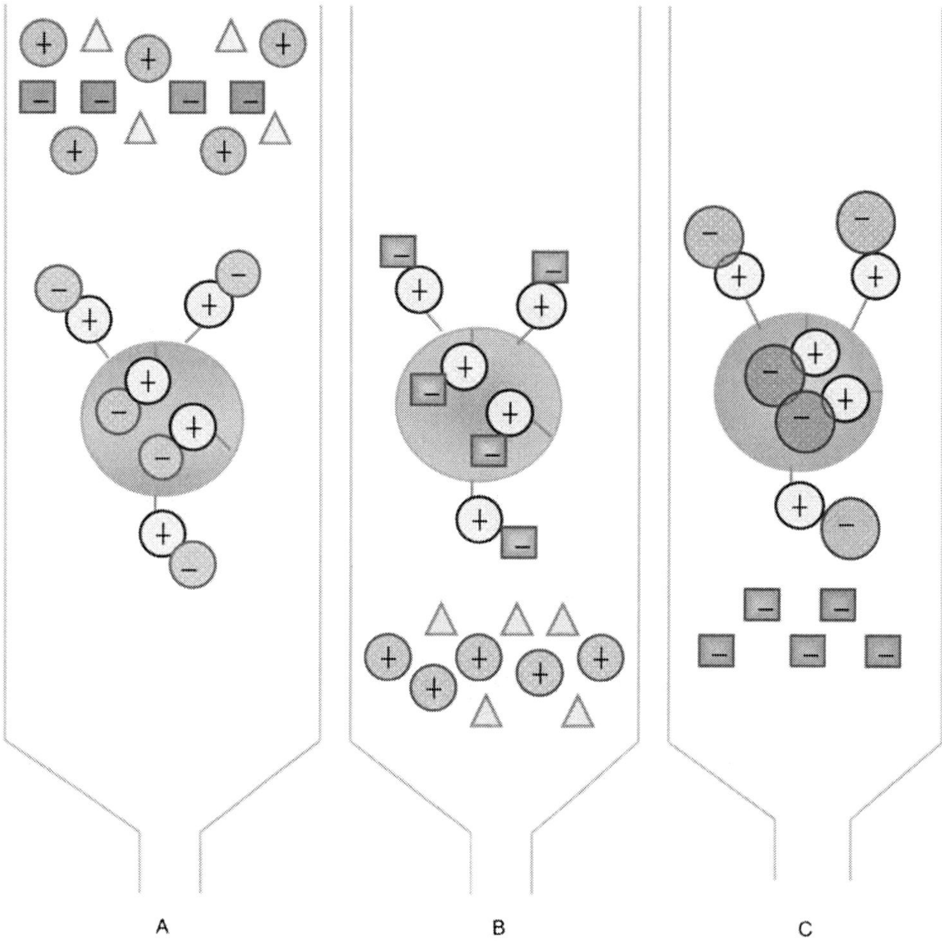

Figure 4.1. Ion Exchange Chromatography.

Ideally, the column is packed with a stationary phase consisting of a synthetic resin that is tagged with ionic functional groups. The insoluble resin material, the matrix, which is positively charged is surrounded by buffer counterions (negatively charged) [A]. When the protein sample is loaded onto the column of stationary phase, the positively charged resin molecules adsorb the negatively charged species of interest, by replacing the buffer counterions. The resin does not bind to the other charged molecules, which are washed away by the buffer [B]. The washing of other protein molecules by the buffer is followed by a short wash of the column with the same negatively charged buffer to remove any unbound protein molecules. The next step is the elution of the protein molecule of interest by use of a strongly negatively charged eluent buffer. This process takes place when the strongly negatively charged species binds preferably to the ion-exchange resin and thereby, displacing the negatively charged protein sample [C]. This purified protein sample is thereby fractionated from its mixture. The aliquots of the sample collected in fractions are pooled up and further used for the next steps of either another purification protocol or analysis of the activity profile and protein content.

Varied ranges of cationic or anionic ion-exchange resin molecules are available commercially to suit the purification of a wide range of proteins (see Section II, Appendix IV B). Many number of eluents may be used for this type of chromatography, however, the use of salt ion buffers suit this purpose. This is in favor because the pH of a buffer determines its ionic strength and can be easily manipulated for the protein of interest.

The strength of binding of the protein sample to the adsorbent depends upon the size of the charge and charge density. The greater the charge or the charge density, the stronger is the interaction. An advanced version of ion-exchange column is chromatofocusing, which allows separation of proteins based on their isoelectric points, by the formation of a pH gradient on the ion-exchange column.

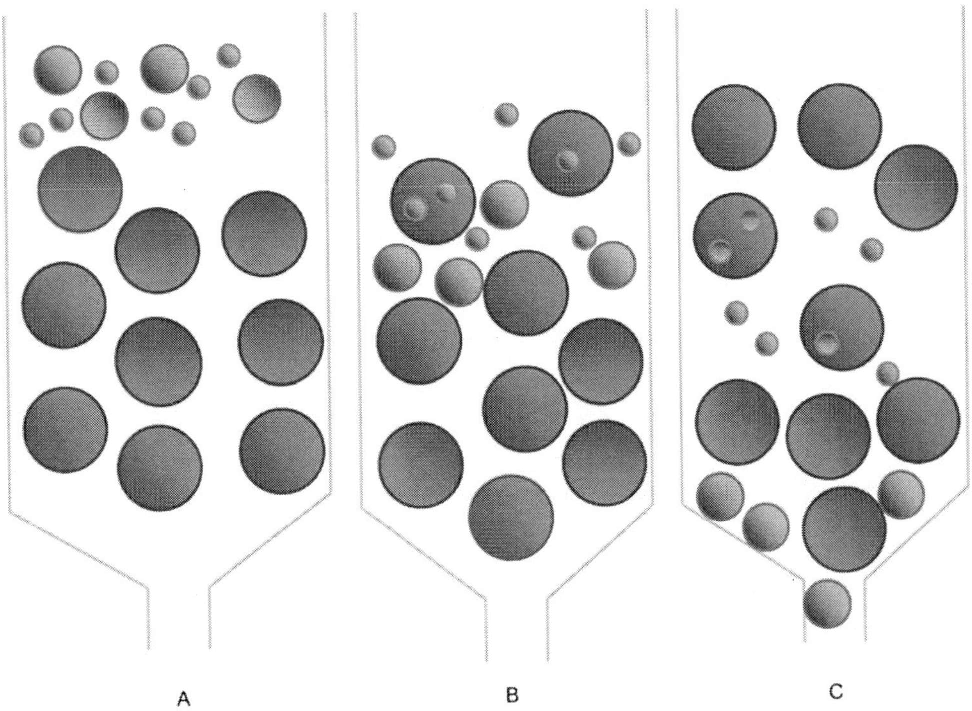

Figure 4.2. Gel filtration chromatography.

4.5.2. Gel Filtration Chromatography

Gel filtration column chromatography is synonymous with molecular sieve chromatography, gel exclusion chromatography, and gel permeation chromatography. Gel filtration chromatography is based on the principle of molecular size of the protein mixtures. This is based on the fact that gels are formed by swelling of the gel matrices and used as stationary phase. Figure 4.2 illustrates the gel filtration column chromatography procedure.

The protein mixture is allowed to pass through the gel column and fractionation takes place according to molecular size and porosity of the gel matrix [A]. The gel matrix has innumerous pores formed by swelling up and protein molecules of smaller sizes are retained in the pores and traverse along the length of the stationary phase [B]. Larger protein molecules do not pass through the pores but flows through the column interspaces and are eluted first [C]. Moderately sized molecules pass through the pores as well as the interspaces and are eluted slowly but prior to smaller proteins. Thus fractionation occurs with continuous solvent flow and the desired protein molecules, which may be small, moderate or large can be purified. The choice of a gel matrix depends upon the MW of the protein which determines the fractionation range of the matrix. It is ideal to choose a gel matrix with a fractionation range that is within the protein MW range (see Section II, Appendix IV A). Gel filtration has further applicability in desalting of proteins and nucleic acids and estimations of MW of proteins using a calibration curve. Some of the terms used in a gel column technique are defined in Box 4.3.

Box 4.3. Terms Used in Gel Filtration Column Chromatography

Water regain: The weight of water taken by 1g of the dry gel. This has been unused recently due to availability of preswollen gel matrices.

Bed volume: The final volume taken up by 1g of dry gel when swollen with water. For preswollen matrices, it is the amount of matrice that is packed in the column.

Void volume: This is the total space surrounding the gel matrix in a packed column. It can be determined by the measure of the volume of solvent required to completely elute a solute that is completely excluded from the gel matrix. Void volume is depicted as V_o.

Elution volume: The volume of eluting buffer necessary to remove a particular solute from a packed column. This is depicted as V_e.

4.5.3. Hydrophobic Interaction Chromatography

HIC is a method for purification of proteins based on their hydrophobicity. Hydrophobic proteins (lipase) are non-ionic and hence exploitation of their hydrophobicity (presence of hydrophobic amino acid residues) to attach to inert matrices containing non-ionic octyl or phenyl groups result in purification. Phenyl Sepharose is less hydrophobic than octyl Sepharose and in the order of hydrophobicity of proteins to be purified, lesser hydrophobic proteins should be treated with octyl Sepharose while more hydrophobic proteins should be treated with phenyl Sepharose. The purification matrix whether octyl or phenyl requires the presence of salts such as ammonium sulfate which increase the affinity of binding of the protein to the matrix. Thereby, a purification protocol by HIC first requires the fractionation of ammonium sulfate and without being dialysed to remove the salt ions, the precipitated protein can be dissolved in buffer (Tris.HCl, pH 8.0) and further be loaded onto the matrix. Generally, HIC using octyl or phenyl Sepharoses require RT for the binding of the protein, while increasing pH reduce/inhibit hydrophobic interactions. Once the column is loaded with the salt-concentrated sample with continuous buffer run to enable protein binding (usually 1-2 h), washing of the column is required to remove any unbound proteins. The column is now ready for elution which has to be carried out at 4°C. A many number of eluents can be used for the elution of the protein. Elution is performed to weaken the hydrophobic interaction between the column matrice and the protein which invariably result in the purified protein to elute out of the column. Eluents for lipases are methanol, isopropanol, ethylene glycol, and detergents such as Triton X-100 and Tweens. Triton X-100 is a good eluent for lipases, however, the activity profile has to be known prior to use. The solvents and detergents have to be used at a concentration well above the protein hydrophobicity so as to displace the lipase enzyme from the matrix. However, removal of detergents is a laborious procedure involving further column purifications and may have a chaotropic effect on the protein.

Use of salts as eluents is also of good consideration and has to be used as a decreasing gradient. The presence of decreasing ionic strength will decrease the binding strength resulting in elution.

Box 4.4. Hofmeister Series of Salt Ions Based upon Decreasing Order of Hydrophobicity

Anions : $PO4^{3-}$ >$SO4^{2-}$ > $CH3COO^{-}$ > Cl^{-} > Br^{-} > $NO3^{-}$ >$ClO4^{-}$ > I^{-} > SCN^{-}
Cations : $NH4^{+}$ > Rb^{+} > K^{+} > Na^{+} > Cs^{+} > Li^{+} > Mg^{2+} > Ca^{2+} > Ba^{2+}

Box 4.5. Preparation of Butyl Toyopearl, a More Hydrophobic Adsorbent

Butyl Toyopearl 650 was prepared from Toyopearl HW-65. 120 ml of Toyopearl HW-65F was washed with 200 ml of water three times and then with 200 ml of N,N-DMF five times on a glass filter. The washed gel, 100g was suspended in 200 ml of of N,N-DMF with stirring. 50 ml of butyl glycidyl ether, 2g of potassium carbonate were added and the mixture was incubated at 70°C for 10 h. The resulting butyl-Toyopearl 650 was washed with 200 ml of N,N-DMF 5 times and then with 200 ml of water 10 times.

Phenyl or octyl Sepharose is supplied preswollen in an ethanolic suspension. Ethanol hinders hydrophobic interaction and hence the column should be washed free of ethanol. After column run, the matrix can be recovered by washing of the column bed thrice or more with distilled water and the matrix can be stored in a 20% ethanol suspension to remove any microbial contaminations till further use. Alternatively, the matrix can be re-equilibrated with buffer for subsequent column runs.

4.5.4. Affinity Chromatography

Affinity chromatography is an excellent method for purification of proteins that display specifity of binding to ligands which are attached to an inert matrix. Affinity columns therefore use the exploitation of the activity of the proteins and their target specificity. A best example of an affinity purification procedure is based on that for glycoproteins. Lipases of eukaryotic origin are essentially glycoproteins, containing mannose residues. Concanavalin A specifically bind to mannose residues and therefore a ConA-Sepharose matrix could be usefully suited for lipase purifications. ConA-Sepharose is supplied as a preswollen matrix from Pharmacia. In running an affinity column, it is essential to know that salts are required for binding efficiency and hence an ammonium sulfate precipitate, similar to HIC lipase sample, can be loaded to the matrix. Salts also prevent non-specific binding to the matrix. Addition of $MnCl_2$ and $CaCl_2$ at 1mM concentration is required for efficient binding of the glycoprotein (lipase) to the ligand (ConA). Since purified proteins require quantification procedures, free sugars should be avoided in the sample. Detergents destabilize the protein structure as well as reduce affinity interactions, for extracellular enzymes, and are needed only for membrane-bound proteins. The affinity procedures for lipase require stronger eluents that have a greater specificity to the ligand bound to the matrix. The mostly specific and preferable eluent is α-D-methyl mannoside (0-0.5 M) which specifically alters the binding capacity of ConA to lipase and in turn binds to ConA, thereby yielding eluted purified lipase. For commercial applications, a combination of mannose/ethylene glycol in a 50:50 (v/v) proportion can be used to elute lipase. This has advantages than α-D-methyl mannoside in terms of costs and large-scale procedures. However, to study crystallographic data or structure-function properties, α-D-methyl mannoside is the only preferred choice. We have also cited a purification strategy as applicable to lipase purification in 4.6 that details practical aspects.

4.5.5. HPLC and FPLC

HPLC is an abbreviation for High Performance Liquid chromatography which is the most preferable form of chromatography in recent research. High pressure, as the name implies, is applied to a column which may be ion-exchange, gel filtration or reverse-phase matrice, which accelerates the sample to be eluted with high resolution and high speed in the order of 15–30 min. than conventional procedures that may take several hours to few days. HPLC has the advantages of resolution and speed, reproducibility, automations, and instantaneous reuse of columns, and yield of high purity proteins and can be used for analytical or preparative purposes. An important parameter in HPLC is the retention time based on which the proteins are analysed, is the time required for maximum elution of the desired protein. The samples that flow through the column under high pressures are detected by differential refractometers or photodetectors or fluorescence detectors. The process is relatively simple to perform, however the complexity of instrumentation needs careful operations. A modified form of HPLC is FPLC which stands for Fast protein liquid chromatography is based upon fast delivery of the protein sample with a flow rate of 1–499 ml/h and operating pressures of 0–40 bar. In HPLC, the flow rate is usually 0.01–10 ml/min and 1–400 bar operating pressure. FPLC can be applied to columns where the sample may be affected in high pressures and similar to HPLC, can be used for a wide range of column applications.

4.6. Planning a Lipase Purification Strategy by Affinity Chromatography

Lipases are hydrophobic proteins and from eukaryotic sources are glycosylated with high percentage of mannose residues. HIC purification methods are on the common run in several laboratories, while it may be ideal to exploit the property of glycosylated lipase proteins by an affinity column method. Affinity purifications work on the principle of specificity of binding to ligands on the inert matrices and would be an efficient method in terms of high purity and yield. Scarce reports on using this purification method are available, and hence, we pertain to the application of affinity column for the purification of an eukaryotic lipase. Consider a microbial lipase from a yeast species, *Candida rugosa*, a lipase model that has been worked out for several years on its characteristics especially for its wide biocatalytic activities.

From the fermentation medium for this yeast lipase, the cells are harvested and the crude enzyme is taken for the study.

Ammonium Sulfate Precipitation

The crude enzyme is first subjected to concentration and fractionation by ammonium sulfate precipitation to 50–80% saturation where the enzyme fraction is obtained. The fractionated enzyme, say at 80% contains maximum lipase activity, is dissolved in a Tris.HCl buffer, pH 8.0. The buffered suspension is then loaded onto a packed column containing the affinity resin.

Packing of Affinity Matrix

The affinity matrix, preswollen ConA-Sepharose is measured to a suitable bed volume lesser than three-fourths of the column (a short and a wide column is preferred) is packed by pouring over a glass rod evenly without allowing any air bubbles. The matrix may be suitably degassed prior to loading. The matrix is allowed to settle by gravity, the whole procedure carried out at 4°C. It is also good to use pre-packed affinity columns that is considerably expensive. The matrix is prevented from drying by use of an equilibration buffer to run through the column continuously for 3–4 bed volumes. The stop cock is then closed allowing a 3 cm buffer layer on top of the column matrix. The column is now ready for loading.

Sample Loading

The lipase suspension is pipetted out onto the top of the column in a continuous stretch without any physical disturbance. Once the loading is complete, the sample layer is allowed to settle for few minutes. The stop cock is now opened and the buffer flow is started. This procedure will enable equal mixing of sample to ensure continuous sample flow. The equilibration buffer is allowed to flow at a constant flow rate, about 2 ml/min standardized earlier using a peristaltic pump.

Purification of Lipase

The lipase sample now binds to the ConA ligand with its mannose residues in the presence of divalent ions such as 1 mM $MnCl_2$ and 1 mM $CaCl_2$ in the equilibration buffer. Usually, a protease inhibitor may also be added, and EDTA should not be added which would otherwise chelate the divalent ions. The ions help in efficient binding of lipase and the unbound proteins are run through the column. Detergents should generally be avoided due to inefficiency of binding of protein with ConA. The column is washed with 2–3 bed volumes of wash buffer to remove unbound proteins.

Elution of Lipase

Elution of lipase from the bound ConA requires the use of an eluent with stronger affinity for ConA that displaces the interactions between mannose residues of lipase which is collected through column fractions. An ideal eluent is α-D-methyl mannoside that yields a pure lipase suited for structure analyses. A commercial eluent is 50:50 v/v mannose/ethylene glycol and can be used on large-scale purification levels. Either eluent is used to obtain purified fractions of lipase collected in a fraction collector at 4°C, pooled and analysed for lipase activity profile and protein quantifications.

An ideal purification should yield high specific activity of the pure protein when compared to the crude enzyme and is a determinant of purity of the protein. Purification yield should also be checked to know an estimate of the purification protocol. The purification analysis is usually tabulated to identify its parameters. The table given below describes the purification profile in complete for a given lipase.

Table 4.2. Lipase Purification Analysis

Method	Total activity (U)	Total protein (mg)	Specific activity (U/mg)	Purification-fold	Yield (%)
Crude enzyme	110.0	55.4	1.98	1.0	100
Ammonium sulphate fraction	106.0	40.2	2.64	1.33	96.4
ConA-Sepharose fraction	72.6	0.6	121.0	61.11	66.0

4.7. Polyacrylamide Gel Electrophoresis (PAGE)

The final step of any purification procedure is the assessment of the purified protein by PAGE with/without sodium dodecyl sulfate (SDS) in an electrophoresis apparatus. Electrophoresis is based on the principle of migration of proteins under the influence of an electric field. Proteins are basically positively charged molecules and in an electric current move towards the cathode. An electrophoresis apparatus is available commercially and it consists of a buffer tank with glass plates holding the gel tightly clamped to the tank with electrodes to allow supply of electric current and with application of a voltage. The gel contains wells as reservoir for protein samples which are prepared with a tracking dye, usually bromophenol blue. The tracking dye allows the monitoring of the PAGE run and aids in completion. Once the PAGE is over, the gels are stained for visualization which appear as bands across the gel. A marker may be provided in another well along with the sample that provides an estimate of the MWs of the proteins to be analysed. Electrophoresed gels are then documented in a gel documentation unit that also applies a densitometric scan of the gel. Thus, the researcher would be able to assess the purity of the protein, its MW, and the densitometric analysis of different protein bands in a preparative gel.

PAGE uses the property of polymerization of acrylamide, with a cross-linking agent, *N,N'*-methylene bisacrylamide in the presence of a catalyst, *N,N,N',N'*-tetramethyl-ethylenediamine, TEMED and an initiator, ammonium persulfate, APS. The polyacrylamide forms a gel inbetween two supports and is placed in the electrophoretic apparatus. Ethanol hinders polymerization and wiping of glass plates with ethanol has to be avoided. The polymerization of the gel with varying compositions yields varying pore sizes of the gel to allow the proteins to migrate through. A high MW protein should be analysed with a lower polyacrylamide concentration that yield gels with larger pore sizes, while a low MW protein should be treated with a high polyacrylamide concentration with lower gel pore sizes. This is to allow rapid migration through the gel matrix. A large number of companies have produced precast polyacrylamide gels of varying % for speed and accuracy in performing a PAGE. Troubleshooting when performing a PAGE run is given in Section II, Appendix III B.

4.7.1. SDS-PAGE

A most commonly and routinely used PAGE analysis is the use of SDS-PAGE to confirm the purity of a protein and analyse its MW. This is suited for monomeric proteins which yield a single band upon PAGE. The principle of a SDS-PAGE is that 1g of protein binds to 1.4g of SDS which equalizes the negative charge on the protein. The flow of the protein is now based only on its MW which attributes this procedure for MW estimations. For an oligomeric protein, the polypeptide chains (subunits) are separated from each other and bind in proportion to SDS with net negative charge remaining the same, and yields more than a single band depending upon the number of monomeric polypeptide chains. A conventional SDS-PAGE includes SDS in the sample buffer, resolving gel, stacking gel and running buffer. The various components used for the preparation of a gel is mentioned in Box 4.6.

Box 4.6. Components of a Polyacrylamide Gel Matrix

All components should be prepared in MilliQ water or atleast DDH_2O.

Acrylamide solution: A 30% stock solution containing 29% acrylamide and 1% bis acrylamide is prepared in water, checked to pH 7.0 and stored in brown bottles to avoid deamination. This stock can be stored in cold for 6 months.

SDS solution: A 10% stock solution can be prepared in water and stored at RT for 6 months. Cold storage will freeze the solution.

Tris buffers: Tris base solution is treated with conc.HCl to pH 8.8 for resolving gels and pH 6.8 for stacking gels at appropriate Molar concentrations. Tris.HCl and Trizma should not be used for buffer preparation, since high salt concentration result in anomalous migration and diffused bands.

TEMED: This catalyses the polymerization of acrylamide and bisacrylamide by free radical formation from APS. TEMED should be the last component added to the resolving or stacking gel mix. TEMED gives an unpleasant odor and has to be tightly recapped immediately after use. TEMED can be stored at RT.

APS: This provides free radicals for polymerization of acrylamide gel. A 10% stock solution is to be prepared prior to preparation of the gel at 100 mg/ml. APS should be always prepared fresh and this has importance in polymerization of the gel, otherwise, would result in non-polymerization. APS has to be stored in a dessicator due to hygroscopic nature of the chemical.

β-mercaptoethanol (BME): This reagent can be stored at RT and is to be added only to the sample buffer mix, prior to running the gel. For a 100µl of sample mix, 5 µl of BME is added freshly.

Glycerol: This solvent is added to the sample buffer at 20% concentration to ensure proper loading of the sample into the wells of the stacking gel, by increasing the density of the sample mix.

Bromophenol blue: This is used at 0.02% concentration to the sample mix, to enable visualization and ensure completion of the gel run by serving as a tracking dye. This can be used as a pinch of dye powder to the sample buffer directly.

Box 4.7. Sample Reloading in a SDS PAGE

Protein samples can be reloaded into the stacking gel wells about 1cm, with 1X sample buffer. The sample buffer contained 60 mM Tris. HCl, pH 6.8, 25% glycerol, 2% SDS, 14.4 mM 2-mercaptoethanol and 0.1% bromophenol blue. 50 µl of protein sample can be reloaded 5 times holding 250 µl of sample. Efficient and sharp bands can be observed by this reloaded protein sample. NaCl may or need not be used in the 1X sample buffer, but also leads to band widening.

4.7.2. Native PAGE

Another version of a PAGE is Native PAGE where the protein is electrophoresed under native conditions, without using SDS which would result in denaturation of proteins having more than one polypeptide chain and β-mercaptoethanol which is used to cleave any S-S bonds in the polypeptide chain (s). Also boiling of sample in sample buffer is to be avoided. This results in retaining of the native conformation of the sample protein and can be electrophoresed. However SDS may be used in sample buffers of monomeric proteins and in running buffers to aid migration. This is an important technique for analysis of proteins which would be further subjected to structural studies.

Box 4.8. Protein Extraction from PAGE Gels

After the PAGE run, the bands of interest were cut with a razor blade and washed thrice, 5 min each, with 2 ml of 250 mM Tris buffer/250 mM EDTA, pH 7.7, in a 25 ml Falcon tube, followed by three rinses of 5 min with DDH_2O. The water was removed with a Pasteur pipette and the gel slices were homogenised with a spatula. 1.0 ml of 20 mM Tris buffer, pH 7.4, containing 0.1% SDS (2:1 buffer : gel v/v) was added. The samples were sonicated for 3 min in an ice-bath with 5-6 passes of 30s.

The sonicated gel was separated from the extraction buffer by addition of 1.5 ml of sample onto a Penesfky column and centrifuged for 10 min at 500xg. A final 1.0 ml volume of protein sample free of gel matrix was obtained. The column was built with a 5 ml syringe filled to 1.0 ml with Sephadex G-25 resin, equilibrated with 20 mM Tris buffer, pH 7.4, containing 0.1% v/v SDS. Prior to sample application, the column was dried by centrifugation. This procedure can be done withTris buffer with/without SDS.

Protein samples free of gel matrix that are not used can be stored at-20°C.

Box 4.9. Zymogram (Activity Ataining) of Lipase

Activity staining is an important technique for the assessment of enzyme activities of proteins in a native gel. Lipase activity staining, in otherwords, a zymogram, is used to identify and check the presence of an active lipase which can be compared alongside a band of the same enzyme. Zymogram of lipase is especially useful when comparing the bands obtained by an SDS-PAGE that has trimeric or tetrameric subunits which can be confirmed to be pure protein with a single activity band.

Native gel is performed and the gel obtained is washed several times to remove any salts and ions. It is then treated with a solution of tributyrin in water, homogenized well and immersed O/N. The active protein forms a clearance zone in the gel indicating lipase activity.

4.7.3. Staining of Gels

Staining of gels is an important prerequisite for detection of proteins after an electrophoretic run. The PAGE gels have to be analysed for elucidation of important characteristics of protein patterns and hence an efficient method of protein detection in gels has to be exploited. Coomassie Brilliant Blue R-250 staining is the most commonly used staining procedures especially of qualitative purposes in confirmation of protein purity and determinations of MWs. This is the routine labscale procedure for detection of proteins in the order of 0.5 μg of protein/cm^{-2} of gel. Coomassie Brilliant Blue R-250 is a dye which binds to protein molecules for easy visualization of gels. This dye is dissolved in a solvent containing 10% methanol and 7% acetic acid which when treated with protein gels O/N yield blue colored bands of reasonable clarity depending upon the protein concentration range. This conventional method has been improved with suitable modifications, such as addition of 25%TCA as fixative prior to staining and use of Coomassie Brilliant Blue G-250. This staining procedure requires a rotary shaking for even binding of the dye and subsequent destaining of the gel. Destaining solution is the same solvent but without the dye. Usually destaining requires thrice or four times washing procedure, with shaking. Eventually the protein bands are only stained with the dye against the gel background. Stained gels can be stored in 7% acetic acid or gel-dried and preserved. A disadvantage of this method is its insensitivity to ng protein detections.

Another method for protein detections that is highly sensitive compared to CBB staining is the silver stain procedure. This is also a common labscale procedure for protein detections in the order of 0.05-0.1 ng protein. However, sensitive it may be it requires time efficiency between consecutive steps that are to be followed. A procedure for silver staining of protein gels is tabulated.

In a silver stain, an important consideration is the time required for development of the protein band and the stop solution is required to arrest the development. Silver stain is a sensitive and a very rapid method for detection of protein bands and is of immense value to biochemists.

Table 4.3. Silver Staining Procedure for Protein Detection in PAGE Gels

Solution	Reagents	Volume (100 ml)	Operation mode	Time
Solution A	MeOH 50% / HOAc 5%		Fix	30 min
Solution B	MeOH 50%		Incubate	15 min
Solution C	Sodium thiosulfate	0.2 g/L (fresh)	Incubate	60s
	MilliQ water		Wash 2 times	2 x 60s
Solution D	Silver nitrate, chilled at 4°C	0.2 g/100ml	Incubate	25 min
	MilliQ water		Wash 2 times	2 x 60s
Solution E	Anh. Sodium carbonate	3 g/100 ml	Develop	Max. 10 min
	Formalin (37% HCHO)	24 μl/100 ml		
Solution F	Sodium EDTA	14 g/L	Stop develop	10 min
	MilliQ water		Wash	2 x 1 min.

4.8. Biochemical Characterization

4.8.1. Thermostability

The thermostability of any given enzyme that is commercially used, such as lipases, is an important parameter for useful industrial applications. Thermal stability depends on the nature of microorganism as well as its source. A many number of thermophilic organisms are isolated from hot springs which impart thermostability to the enzyme under consideration. For example, *Thermophilus* sp., including *Thermophilus aquaticus* from which *Taq* polymerase is obtained, are obtained from natural hot springs and exhibit thermostability to the order of 90-100°C as their temperature optimum. The thermostability of enzymes is a consequence of the proper conformations of the protein and unique DNA sequences that contribute to its thermostability. Thermostability is a key feature for industrial enzymes because of the high temperatures applied for biocatalytic applications to produce synthetic products. Thermal stability may be characterized by determinations of lipase activity of a mixture of equal volumes of lipase and a suitable buffer incubated at various temperatures at

30 min. time intervals upto 2-3 h. This enables the researcher to identify the most applicable temperature for the enzyme catalysis.

4.8.2. Substrate Specificity

The natural substrates for lipases are the triglycerides such as triolein, tripalmitin and tristearin. However, the number of sources available for lipase isolation has evolved in acclimatization of the enzyme to various substrates exhibiting substrate specificity. A lipase from the same species, but from different sources will have rarely the same specificity and usually a different specificity towards their substrates. Hence, a study of the substrate specificity of a lipase is essential to understand its characteristic features. Various triglycerides from triacetin to triolein may be studied for any given lipase which is also suitable for commercial applications. *Penicillium camembertii* lipase exhibits substrate specificity towards DAG and MAG but not TAG.

4.8.3. Fatty Acid Specificity

Any lipase source, say *Candida rugosa* or *Pseudomonas aeruginosa* exhibit characteristic fatty acid specificity or chain length specificity of the fatty acid moieties present in triglycerides. Triolein contains oleic acid as the fatty acid moiety and is a C18 molecule. Similarly, C2-C18 fatty acids may be tested for identification of fatty acid specificity. *Geotrichum candidum* lipase is a unique lipase in the sense that it targets only the TAG with fatty acid moieties containing oleic acid residues that have a double bond at the *cis*-9 position of the fatty acid.

4.8.4. Positional Specificity

The most unique feature of any lipase is its positional specificity (see Chapter 1, unit 1.4). The determination of positional specificity of a lipase is the first step towards biocatalytic applications. We hereby describe the protocol for identification of the positional specificity of a *Pseudomonas* lipase.

Box 4.10. Determination of Positional Specificity

100 mM triolein, 9 ml of Tris.HCl buffer, pH 7.0 and 2 ml of lipase were shaken at 200 rpm at 37°C. At different time intervals, 0, 3, 6, 9, and 12 h, reaction products were extracted with n-hexane or diethyl ether. TLC was performed with pre-coated Silical gel plate using a solvent mixture of chloroform and acetone in the ratio of 96:4 v/v. The samples were run in different lanes with standards for triolen products. Iodine vapor was used for detection of the reaction products. The triolein products were 1(3) monoolein, 2-monoolein, 1,2 (2,3-) diolein and oleic acid. The reaction products that correlated with the triolein products were identified and characterized for positional specificity.

4.8.5. Electrophoretic Mobility

Proteins possess characteristic electrophoretic mobilities that enable identification of their structural conformations. Lipases with inactive forms have a closed structure while those with open conformations are active. This discrimination of open and closed forms can be detected in a native isoelectric focusing gel with their differences in migration based on their pI.

4.8.6. Structural Features

Lipases possess unique structural features that are characteristic of independent lipase families with strong correlation to its members. The most significant structural features are its active site, lid structure and hydrophobicity. An important aspect of the lipase active site is that it is buried in a hydrophobic pocket with a closed lid domain. Ser-His-Asp forms the active site similar to Ser-proteases. In case of *Geotrichum candidum* lipase, Asp is replaced by a Glu residue. The lid structure uncovers the active site in presence of a substrate and the reaction takes place at the interface. To consider another important feature is its hydrophobicity and a consensus sequence. The hydrophobic residues contribute to the overall native conformation of lipase, composed mainly of Trp residues. The consensus sequence is G-X-S-X-G, where X may be any other residue. Generally lipases range from 20-60 kDa and are usually monomeric. Still, oligomeric structures have also been investigated. Chemical inactivation studies had earlier contributed to identification of essential residues of proteins under investigation. PMSF and DIPF are Ser inhibitors. IAA inhibition provides a clue to the presence of a -SH bond, while DEPC is a His inhibitor. 2-mercaptoethanol identifies a S-S bond while NBS inhibits Trp. Structure analyses of lipases are innumerous and contribute towards the overall identity of lipase structures, using circular dichroism, fluorescence spectra, NMR, X-ray crystallography, IEF, 2-D PAGE and bioinformatics tools. In view of the large data of lipase structures available, lipase structure-function relationships can be studied to enable further insights and resolution of complexity of lipase structural features.

4.8.7. Chemical Modification

An aspect of structural feature of lipase is their chemical modifications for commercial applicabilities. Lipases have been modified with Tyr residues and acylating agents to enhance their properties such as esterification abilities. Surfactant-coated lipases are immensely used in nonaqueous media for various bioctalytic activities. This is an interesting area of research to those who are working on lipases which provides many avenues for successful commercial applications.

Self evaluation

(1) Column chromatography based on protein sizes is

(a) Gel filtration (b) Affinity (c) GC (d) IEF

(2) DEAE Sepharose is
(a) Anion exchanger (b) Cation exchanger (c) agarose matrix (d) a & c

(3) FPLC is applied with
(a) high pressure (b) fast flow rate (c) both (d) none

(4) TLC is used for lipase
(a) Positional specificity (b) purification (c) PAGE (d) Staining of gel

(5) Zymogram can be used for
(a) native enzymes (b) lipase (c) both (d) none

Answers

1a 2a 3c 4a 5c

Chapter 5

LIPASE STRUCTURE

5.1. INTRODUCTION

Lipases display characteristic features that are unique to its own class of enzymes, the triacylglycerol hydrolases, depending upon the nature of sources and the conditions of suitable isolation methods. A number of techniques and methods to characterize the lipase enzyme are largely available and since the 1990s, several efforts have been taken to identify the structural features of lipases. The RCSB PDB (http://www.rcsb.org/pdb/) reports 130 lipase structure hits till date.

The structure prediction tools from Bioinformatics Labs have also contributed to the vast structure databases. An understanding of the structure of lipases enable structure-function relationships and beyond the proteome and promote better modulations of the enzyme suitable for useful applications as biocatalysts, an area known as protein engineering. In this context, we describe the structural tools that are currently used for identification of protein structures and detailed structural features of lipases so far studied.

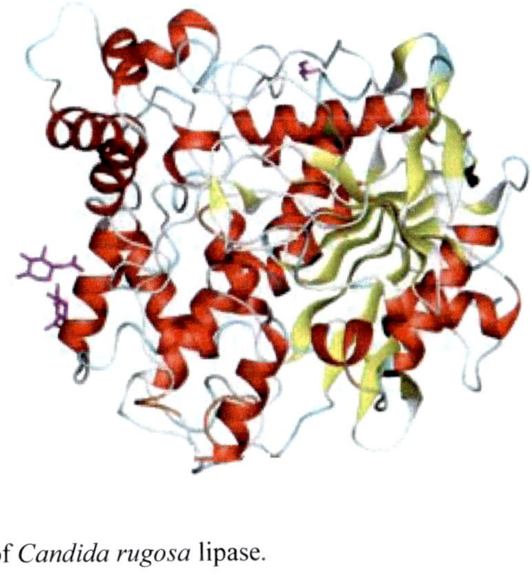

Figure 5.1. 3-D structure of *Candida rugosa* lipase.

5.2. Structure Analysis

5.2.1. Circular Dichroism

Circular Dichroism (CD) is a potential tool for the structure analysis of proteins and the significant rise of research in structural biology is attributed to this method of secondary structure analysis. With elucidations of secondary structure and the prime parameters of alpha helices, beta sheets, random coil structures, beta turns, the CD spectroscopy is the best method of choice for protein structure prediction. Various online sources aid in the structure prediction using CD spectra from different laboratories' Web servers. A general understanding of CD is essential and we describe the practical perspectives of CD analysis. First, we provide the approaches for making a CD spectral analysis and we consider the different parameters that can be studied using CD.

Box 5.1. Circular Dichroism

When plane polarized light passes through an optically active solution, the left and right circularly polarized components of the plane polarized light are absorbed by different amounts. When these components are recombined they appear as elliptically polarized light, and the ellipticity is defined as θ. Circular Dichroism is a result of the absorption of linearly polarized light by the chiral components and is measured in ellipticity values.

Generally, CD spectra are carried out in the far-uv and near-uv ranges for an ideal protocol. CD spectra require 20–200 μl of protein solution in the far-uv wavelength in the order of 0.05–1.0 mg/ml concentration dissolved in a buffer that does not have an absorbance capacity in this wavelength. Since most structural data are obtained in the lower uv-range, CD signals may be masked by the buffer absorptivity. 10 mM potassium phosphate is an ideal buffer system. HCl should not be used in the buffer for any adjustment, while mixing proper ratios of monobasic and dibasic phosphates will ensure transparency. Counter ions are generally not required, if they should be used, NaCl should be avoided. Sodium sulfate or sodium fluoride may be used as counterions in the buffer system. Absorbance of the sample is a critical parameter for the CD spectrum. A sample OD of 1.0 may deter accuracy, and should therefore be diluted. Near-uv characteristics require 1 ml of protein solution with an A_{280} of 0.5-1.0 corresponding to 0.25 to 2.0 mg/ml protein. Pathlength of the CD cell smaller than 0.1 mm or 0.05 mm decrease the buffer (solvent) absorbance, which require higher protein concentrations of 1 mg/ml. It is also necessary that protein samples are free of particulates by filtration through a 0.2 or 0.45 μm filter.

Proteins should be atleast of 95% purity as determined from a mass spectrum of whole proteins and peptide fragments. Nucleic acid contaminants should be eliminated prior to sample preparation, monitored using absorption spectrum. Absorption spectrum of proteins should be of the order of atleast $A_{280}/A_{260} = 1.7$, while the nucleic acid ratio should not be above 0.6, which depend upon amino acid and base compositions. Elimination of nucleic acids should be done while obtaining the cell lysates by addition of nucleases prior to purification of the protein. In case of purification of His-tagged proteins, removal of imidazole and chloride ions in ion-exchange columns should be performed by dialysis or gel permeation, since they interfere with absorbance in the far-uv range. Artefacts in the CD spectrum due to light scattering may appear with protein samples containing infiltrates. Storage of protein samples that are to be used for a CD spectral analysis is of prime consideration. Buffers and protective agents may be employed to ensure the stability of the protein during storage. Buffers of low ionic strengths with appropriate pKa values close to that of the protein can be used with advantages that they disperse the surface charges on the protein sample while preservation. Protective agents such as ammonium sulfate, osmolytes such as proline, β-alanine betaine, glycerol, sucrose, are generally used. Addition of 50% glycerol to the protein and storage at -20°C renders sample stability without freezing. 1 mM EDTA absorbs significantly below 200 nm and is an effective storage additive. However, the pathlength cell used should be 0.02 cm, while the EDTA does not hinder with the CD spectrum. Zwitterionic buffers such as HEPES, MOPS, MES, PIPES generally are regarded as safe protein solvents at lower concentrations and absorb below 200 nm for detailed far-uv CD studies. 8 M urea and 6 M GdmCl are routine components in denaturation of proteins and are useful indicators in CD investigations. This high concentration result in unreliable CD data below 210 nm, while changes in CD signals at 222 or 225 nm can be used to assess the unfolding of the protein.

Far-uv is characteristic of secondary structures such as Alpha-helix, beta sheet and random coils. Spectrum is analysed at 190–250 nm in a CD cell of defined pathlength, known aliquot of protein concentration and monitoring the ellipticity in degrees, using a spectropolarimeter. Near-uv spectrum is in the range of 250–350 nm is characteristic of Phe, Tyr and Trp residues (Figure 5.2).

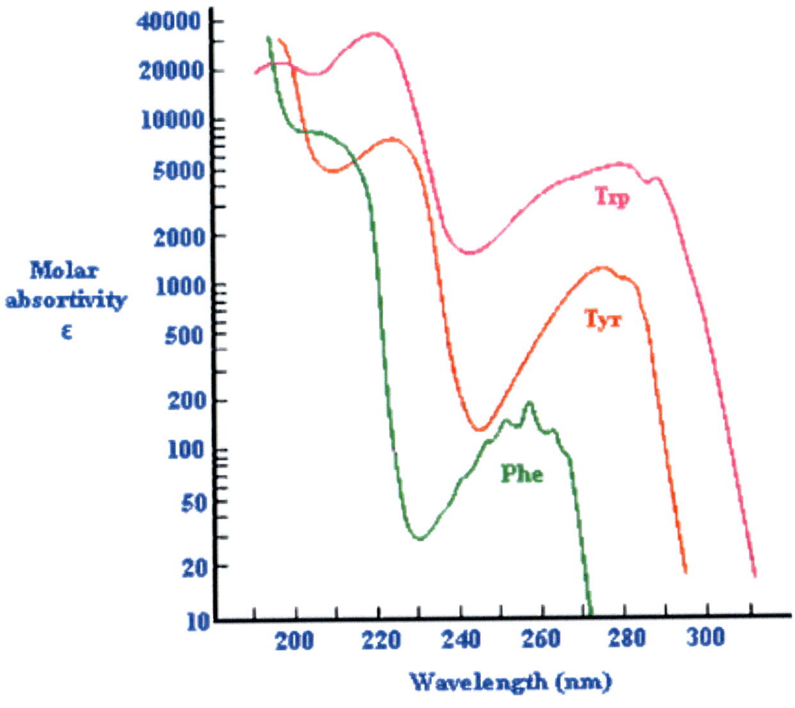

Figure 5.2. Near-uv spectrum of aromatic amino acid residues.

CD spectra can be used to analyze protein conformations, their secondary and tertiary structures, cofactor binding sites, including protein folding. Secondary structural features of proteins, alpha-helices, beta sheets and random coil structures are essentially derived from CD analysis. Synchrotron radiation CD (SRCD) can be used to obtain far-uv ranges of 170 nm and below with reliable estimates of secondary structures. Tertiary structures can be determined by a near-uv spectrum and is indicative of aromatic amino acid residues. Trp, Tyr and Phe have characteristic wavelength profile in this uv range. Trp shows a peak close to 290 nm with a fine structure between 290 and 305 nm. Tyr exhibits a peak between 275 and 282 nm and Phe shows sharp bands at 255 and 270 nm. Cofactors in free solution show little CD signals while in binding with proteins exhibit chirality and hence CD signals are excellent indicators of integrity of cofactor-binding site. Conformational changes in proteins can be effectively monitored in a CD spectrum with binding of ligands to large protein structures. CD is also invaluable for structure determinations of peptides where X-diffraction analyses are difficult. Peptides that switch conformations from α-helix to β-sheets and helical hairpin to a coiled coil can be assessed by CD measurements. Rate of acquisition of secondary and tertiary structures determine the fold of a protein. Continuous or stopped-flow CD methods detect ms or sub ms time-scale levels and are useful experimental tools in protein folding. CD analysis is preferentially employed in denaturation studies of proteins under high temperature variations. It is usually preferable to study the thermal denaturation of proteins to characterize the protein conformations in such states. Far-uv CD spectra allow studies with increasing temperature profiles and monitor secondary structures. The pattern of these spectra indicates structural changes and is characteristic of each protein and enables sufficient data for derivation of thermal stabilities. Also, far-uv spectra of proteins with urea and GdmCl elicit suitable structural information.

CD spectra of unknown proteins can be determined with reference to polylysine spectra. Polylysine is a typical chiral molecule and exhibits optical ellipticity and reference to this protein CD spectrum can interpret structural aspects of an unknown protein.

CD-tool and CD-Pro are software packages that integrate CD spectral analysis and data acquisition and streamline the data analysis. A number of CD analysis softwares are available online for analysis of CD spectral data.

Box 5.2. Far-UV Spectra Characteristics in Polylysine $(Lys)_n$

α-helix:	negative at 222 nm is red-shifted (n → π*)
β-sheet:	negative at 218 nm (π → π*)
	positive at 196 nm (n → π*)
Random coil:	positive at 212 nm (π → π*)
	negative at 195 nm (n → π*)
Conformations:	α-helix at pH 10.8
	β-sheet at pH 11.1 after heating to 52°C and recooling
	random coil at pH 7.0

5.2.2. Fluorescence Spectroscopy

Fluorescence spectra are characteristic of structural features for protein structural perspectives. Fluorescence spectra result from the absorbance as well as the emission wavelengths of proteins of specific interests. This is based on the principle that some molecules are excited during absorbance of light at a specific wavelength. Such molecules do not remain in the excited state for long and tend to reach the ground state by emission of polychromatic light of longer wavelengths, producing a fluorescence spectrum. This spectral analysis is indicative of structural aspects for a given chemical entity and could target at the identification of structural characteristics similar to other spectroscopic methods such as IR, NMR. In case of lipases, there has been strong evidence for the presence of Trp residues in the 'lid' structure that covers or overlaps the active site and the hydrophobic pocket. Trp is an excitable molecule and emits characteristic fluorescence. Trp residues provide intrinsic fluorescent probes for protein structure allowing detailed studies on the conformational dynamics. The fluorescence lifetimes of Trp depend on its microenvironment, allowing further insight into conformational status of the protein under consideration. Steady state fluorescence measurements and time-resolved fluorescence spectroscopy monitor the changes in the fluorescence behavior. Fluorescence anisotrophy reports on the conformation as well as the state of association of the protein. Hence, fluorescent spectra can provide a wealth of information on the properties of lipases due to the features of Trp residues. Apart from Trp, Tyr residues also contribute to fluorescent spectral analysis. Therefore, lipases that contain Trp and Tyr could be enabled for investigations of their structural analyses by fluorescence analytical methods. We hereby report the practical approaches for performing a fluorescence analysis and the features of Trp in lipases studied earlier.

A spectrofluorometer provides the light source for excitation and detects and records the emission spectrum by the protein sample. Excitation wavelengths may be 280 nm or 295 nm to include Trp and Tyr contributions. This eliminates contributions from other amino acid residues. The excitation wavelength at 280 or 295 nm results in fluorescence signals that reflect the global conformational changes of the tertiary structure of proteins. Emission spectra are generally obtained between 310–400 nm. This range predicts whether a Trp residue is present within or outside the hydrophobic pocket. Spectra at 335 nm is characteristic of Trp well buried in the hydrophobic core, while, spectra at 350 nm is characteristic of Trp exposed to aqueous solvent. If a protein containing a single tryptophan in its 'hydrophobic' core is denatured with increasing temperature, a red-shift emission spectrum will appear. This is due to the exposure of the tryptophan to an aqueous environment as opposed to a hydrophobic protein interior.

Chromobacterium viscosum lipase (CVL) contains 3 Trp and 10 Tyr residues. Temperature effects on the fluorescence emission spectra of CVL indicated that at low temperatures, it retains its activity, while increasing temperature leads to a shift in the emission maximum. A red-shift is observed with increase in the polarity indicating temperature-dependence of the Trp. Flourescence of tyrosyl residues presents a less structured state and are blue-shifted. Based on fluorescence data of *Humicola lanuginosa* lipase (HLL), it has been inferred that HLL adapts the open conformation in its substrate, triacetin, with the exposure of the active site by α-helical surface loop and also that Trp89 contained in the lid played an important role in the structural stability of HLL.

5.3. Protein Engineering

The versatility of the many varied catalytic properties of lipases has contributed to the development of protein engineering technologies to yield industrial and value-added products of commercial significance. Lipase protein engineering has evolved since the 1980s and has arisen from the structural elucidations of lipases, which have been deposited in the protein data bank, since its inception. Microbial lipases have been investigated for their characteristic structural properties that impart their biocatalytic activities. The first example on the protein engineering of lipase was from *Pseudomonas mendocina* lipase. Early studies on protein engineering was based only on sequence information and with the advent of sophisticated techniques and instrumentation, crystallography softwares and bioinformatics tools, lipase structure studies have evinced keen interests from among lipase researchers. European Consortium on the structural prediction of enzymes has contributed to the enormous funding on structure investigations. This has prompted the development of protein design suitable for commercial applications targeting at their specificities and structural chemistry. For example, while lipase acts on triacylglycerol esters, it can be made possible for the protein engineering of lipase to target its specificity towards cholesteryl esters and cholesterol to aid in the hydrolysis or degradation of the cholesterol molecules. This is applicable in the medical sciences where it has been yet difficult to reduce cholesterol content in the metabolic status of the subjects. Likewise, protein engineering could result in the productivity of newer molecules of significance for exploitation in suitable areas of interests. Inspite of vast structural information on lipases, efforts are still required to accomplish the structural determinations of many of the lipases that have been extensively studied in basic as well as applied research. A number of lipases of microbial origin have been isolated, a number of them purified and characterized, but with only a few studied for their structural characteristics and structure-function relationships. Site-directed mutagenesis is another important technique in the protein engineering of enzymes or proteins wherein an amino acid may be mutated to another amino acid, which significantly alters the specificity profile of the original protein molecule. The alteration of enzyme properties attributes the enhancement of its biocatalytic efficiency or any structural property. Protein engineering of lipases can be addressed in terms of its variants in active site, substrate, chain length, positional and enantio-specificities, thermal stabilization, stability towards proteases, detergent and surfactant compatibility, lid function and calcium binding properties.

5.4. Lipase Structure

Since the development of sequencing techniques, sophisticated methods of determination of protein structures by X-ray diffraction, 2-D NMR, and availability of bioinformatics databases specific for protein 3-dimensional structures, motifs, patterns, phylogenetic analyses, consortiums of protein structure elucidations have contributed to an increase in rapid structure identification of many different proteins in many different areas of biological sciences. With regard to lipases, the first microbial structures studied from crystal analysis were that of *Rhizomucor miehei* (1990), *Geotrichum candidum* (1991) in which mixed β-pleated sheets were identified and with their active sites buried. In *Rhizomucor miehei* lipase

(RmL), a short helical lid structure which opens up in presence of an oil-water interface was described and the phenomenon of 'interfacial activation' was postulated, while a conformational change in presence of lipid substrates was also postulated with the latter. The interfacial activation of lipase determines the 'closed' and 'open' forms which are related to the inactive and active conformations of the lipase.

The first 3-dimensional lipase structure from a microbial source was elucidated from *Rhizomucor miehei* in its active form by Zygmunt.S.Derewenda, Urszula Derewenda and Guy G.Dodson in 1992, followed by *Rhizopus niveus* lipase structure in 1993 and *Candida rugosa* also in 1993. The first bacterial lipase structure reported was *Pseudomonas glumae* lipase in 1993. *Staphylococcus hyicus* lipase structure was determined in 1995.

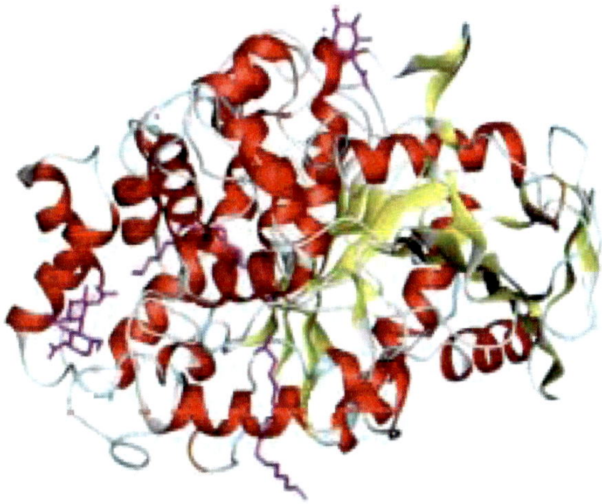

A. open (active)

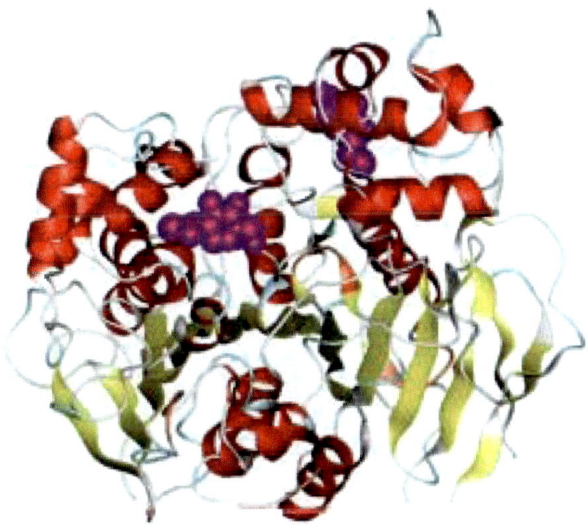

B. closed (inactive)

Figure 5.3. 3-D structures of *Candida rugosa* in open and closed forms.

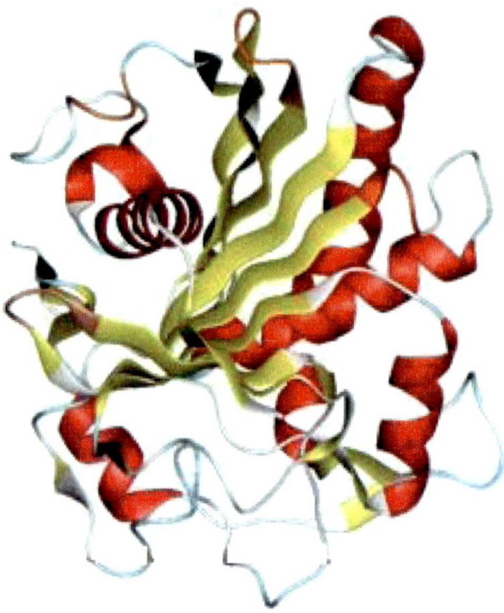

Figure 5.4. 3-D structure of RmL.

5.4.1. Structure of RmL

A single polypeptide chain of 269 residues is folded into a singly wound β-sheet domain with predominantly parallel strands, connected by hairpins, loops and helical segments. The loops are right-handed in which the central sheet is asymmetric and the connecting fragments are located on one side of the sheet. An N-terminal helix supports the distal side of the sheet. Disulfide bonds 29–268, 40–43 and 235–244 stabilize the lipase molecule. Four *cis*-peptide bonds precede proline residues. The active site structure is a constellation of His257, Asp203 and Ser144, similar to the Ser proteinases. Tyr260 is the fourth residue of the active center. This catalytic site is surrounded by an amphipathic helix 85-91 lid, held by hydrophobic interactions.

5.4.2. Open Conformation of *Pseudomonas* Lipase

The most significant feature observed in lipases is the α/β-hydrolase fold. We describe the global fold, active site cleft and the oxyanion hole structure of *Pseudomonas* lipases, with reference to *Pseudomonas cepacia* (PCL), *Pseudomonas glumae* (PGL) and *Chromobacterium viscosum* (CVL).

Alpha/beta Hydrolase Fold

Many features of the global fold are well conserved in the *Pseudomonas* lipases. The central core conforms to the β strands β3–β8 of the α/β-hydrolase fold. The first strand of *Pseudomonas* lipase is named β3 and follows consecutively. Active site Ser87 lies at the C-terminal β5 strand. Mainchain takes a sharp turn at Gly110, located at the C-terminus β6. Strand β7 is the longest strand in the β sheet. Asp264 is part of a loop which follows strand β7 and the triad His286 is located in a loop following β8. There are 11 α-helices corresponding to the conserved helices A-F of the global fold: α1=A, α2=B, α3=C, α7=D, α10=E and α11=F.

Active Site Cleft

The active site is comprised of Ser87, His286 and Asp264 which form a number of H-bonds. The active site serine lies at the bottom of a cleft in the protein and is exposed to the solvent. The cleft measures 10 x 25 A° across and 15 A° deep. The floor of the cleft is U-shaped and at the bottom, near Ser87 is 4 A° wide. The walls of the cleft surrounding the active site are formed primarily by hydrophobic residues and the cleft is bordered by helices α4, α5 and α9 and by residues 17–29.

Oxyanion Hole

The amide N-atoms of Gln88 and Leu17 form an oxyanion hole at the C-terminus strand β3 and the loop of the strand β4. Arg61 is located at the center of a H-bonding network that stabilizes the region around the oxyanion hole.

5.4.3. Disulfide Bond in *Pseudomonas aeruginosa* Lipase

PAL consists of two Cysteine residues which form an intramolecular S-S bond, which stabilizes the active conformation of the lipase, but is however not required for its folding and for the interaction with its foldase. This lipase stabilization occurs during and after translocation of the Type II secretion pathway.

5.4.4. Structural Features of *Penicillium expansum* Lipase

PEL mature peptide residues 130-136 GHSLGGA are highly conserved with four Cys residues Cys25, Cys220, Cys228, Cys254, suggesting the existence of two possible S-S bonds. However, this lipase lacks N-glycosylation site. A high percentage of hydrophobic residues of the order of 51.4% in the N-terminal region has been observed as compared to 31.4% for RmL, 38.3% for *Rhizopus delemar* and 35.8% for *Penicillium camembertii* lipases.

Self-evaluation

(1) A structure analysis method is
(a) X-ray diffraction (b) NMR (c) Circular dichroism (d) all the above
(2) Protein structure database is
(a) PDB (b) NCBI (c) DDBJ (d) KEGG
(3) CD spectra give information on
(a) alpha helix (b) thermal stability (c) aromatic amino acids (d) all the above
(4) Fluorescence spectra is characteristic of
(a) Trp (b) Tyr (c) both (d) none

(5) Lipase structure is
(a) Ser-His-Asp (b) oxyanion hole (c) α/β fold (d) all the above

Answers

1d 2a 3d 4c 5d

Chapter 6

MOLECULAR CLONING AND OVEREXPRESSION

6.1. INTRODUCTION

Genetic engineering is a major discipline in Biotechnology for molecular cloning of enzymes, proteins of interests for research, analytical and commercial prospects. Molecular cloning is an important tool for basic and applied researchers for production of recombinant proteins, sequence information, desired phenotypes, analysis of genetic complements, identifications of gene aberrations in congenital and other genetic disorders, in forensics, gene therapy and DNA vaccines. An approach to produce recombinant enzymes of commercial significance is an essential step for the overproduction of industrial enzymes. An attempt to clone lipase gene is of considerable interest and large amount of literature is available. Still, developments in advanced gene cloning methods and designs of suitably engineered vectors promote further insights into more cloning procedures and establish successful cloning strategies.

Molecular cloning involves the incision of the target gene with restriction endonucleases and clone it into a vector that may be a bacterial plasmid or phage and the clone is transformed into a desired host strain, *E.coli* or yeast. A large number of restriction enzymes have been identified that serve as the cutting tools and with sequence specificities. Restriction enzymes of the Type II recognize a specific DNA sequence and cleave in a predictable and specific manner. These are the commonly used endonucleases in molecular cloning and they act on palindromic DNA sequences and may produce blunt or cohesive DNA ends according to their site of cleavage known as restriction sites. Restriction enzymes are named according to its source of the strain from which it is isolated.

Box 6.1. Palindromic Sequence

Palindromic sequence is a DNA sequence that reads the same sequence in 5′→3′ of both strands. For example, EcoRI recognizes G↓AATTC DNA sequence. This sequence is
CTTAA↑G
complementary as well as palindromic.

Box 6.2. Some Common Restriction Enzymes

BamH I	G↓GATCC
EcoR I	G↓AATTC
Hind III	A↓AGCTT
Kpn I	GGTAC↓C
Not I	GC↓GGCCGC
Pst I	CTGCA↓G
Xho I	C↓TCGAG

Box 6.3. Methylase

Methylase is a bacterial enzyme that adds methyl groups to the site-specific DNA sequence recognized by their restriction enzymes. This prevents the attack of the host restriction enzymes by this DNA-modification system and protects its native DNA.

The desired or target gene is cut at its 5′- and 3′-ends by its sequence specific restriction enzymes. Vectors may generally be plasmids which are extrachromosomal elements and capable of independent replication, usually not requiring the host machinery with introduction of promoter sequences and bacterial or yeast origin of replication sequences.

Box 6.4. Plasmids

Plasmids are small extrachromosomal elements, of 2–200 kb, present in multiple copies and capable of autonomous replication with an origin of replication, however requiring the host polymerases and other replication machinery. This is a primitive concept of earlier plasmids such as pBR322.

pBR322 is 4361 bp in length and contains the replicon *rep* responsible for the replication of plasmid, *rop* gene coding for the Rop protein, which promotes conversion of the unstable RNA I-RNA II complex to a stable complex and decrease copy number, *bla* gene and *tet* gene.

Bla or amp^r gene encodes β-lactamase that degrades penicillin antibiotics (ampicillin); tet A gene encodes a transmembrane pump to remove tetracycline from the cell.

Many number of engineered vectors tailor-made for specific purposes are available and is a topic of specific interest in most research labs. The vector DNA is also cut with the same restriction endonucleases to produce complementary blunt (even fragment) or cohesive (protruding fragment) terminii. The complementary ends are then ligated by a DNA ligase. The ligation procedure involves the joining of the vector (plasmid) and the gene sequence. It is usually desired that ligation kits are used for simplicity and avoid mismatched ligations during incubation. Many standard Companies supply ligation kits for such purposes. It is also advisable to dephosphorylate the vector ends to prevent religation of vector DNA ends. A simple vector dephosphorylation protocol is available from Promega. The ligated vector is now ready for transformation into suitable host strains. Host strains are innumerable and are also supplied with the vectors such as yeast strain for pESC yeast vectors from Stratagene.

E.coli BL21, *E.coli* DH5α are on the common use. This is a simple methodology for a basic cloning strategy. A most important criteria is that the vector should possess suitable markers for identification of transformants such as antibiotic resistance Amp^r, Tet^r gene sequences or a *lac* gene which produces white recombinant colonies against non-transformed blue colonies using IPTG/X-gal. This is an essential and critical step towards successful cloning.

Box 6.5. Applications of rDNA Technology

Recombinant protein production, GMOs, transgenics, DNA Fingerprinting, Plant and Animal breeding, Medical diagnosis, Gene therapy

With this view in mind, we proceed to describe the essential gene cloning techniques, Polymerase chain reaction (PCR) for gene amplification and an RT-PCR cloning strategy for lipase cloning from a microbial source. An analysis of vectors of bacterial, phage and yeast origin are also described, and finally a few practical considerations in gene cloning techniques are noted.

6.2. VECTORS

Vectors are generally carrier sequences that carry a DNA insert for replication and propagation in a host. Earlier, naturally occurring plasmids were used as vectors which contained restriction sites and antibiotic resistance marker and were capable of replication in *E.coli*. Phage vectors were later the choice by infection and formation of plaques, and for ssDNA replication. With the demand for multifunctional cloning, vectors have been engineered to suit various requirements for every cloning laboratory. Promoter sequences, phage polymerases, multiple cloning sites, bacterial, phage and yeast origin of replication and lacZ' sequences have been incorporated to facilitate cloning and expression in varied hosts. It is therefore important to discuss the many different types of vectors which are derived from plasmid, bacteriophage, phagemid, cosmid, yeast, bacterial and yeast artificial chromosomes (BACs and YACs).

Box 6.6. Multiple Cloning Site

Multiple Cloning Site (MCS) is a region of DNA sequence where there are a number of restriction sites for the purpose of multiple cloning. The many number of restriction sites facilitate flexible use of restriction enzymes that may be available in the gene sequence of interest.

```
 EcoR I                                      Not I          Spe I      Cla I
-GAA TTC AAC CCT CAC TAA AGG GCG GCC GCA CTA GTA TCG ATG GAT

                                                  Sac I      Pac I
TAC AAG GAT GAC GAC GAT AAG ATC TGA GCT CTT AAT TAA-
```

Plasmids are the simplest form of vectors and an early version is pBR322. Plasmids are the choice for simple cloning in an *E.coli* host and are in practice for teaching.

Box 6.7. Insertional Inactivation

Plasmids containing antibiotic resistance genes such as amp^r and tet^r allow simpler identification of recombinant plasmids. Any insert incorporated into the amp^r gene or the tet^r gene sequences would result in inactivation of the resistance to the antibiotics. When the recombinant plasmids are allowed to grow in ampicillin plates, DNA insert would make the plasmid sensitive to ampicillin and similarly with tetracycline plates. This is known as insertional inactivation. Since colonies are susceptible to amp or tet, a replica plating has to be made prior to antibiotic selection.

This has however been a conventional screening method and has largely been replaced by blue/white screening that requires a lac gene coding for β-galactosidase.

Engineered versions have an MCS, lacZ′ gene, Ab^r marker and an origin of replication More recently, plasmids have been used as expression vectors for expression of recombinant genes without requiring host machinery. Transcription of lacZ′ gene produces a fusion protein that contains the N-terminal sequence of lacZ′ and the cloned insert sequence. This enables the fusion protein to be purified by His-tag (a series of His residues) column containing Ni ions.

pUC 18 and pUC19 vectors contain polycloning sites arranged in opposite orientations. pUC plasmids lack the *rop* gene which is normally located close to the origin of DNA replication and is involved in the control of copy number. As a result, these plasmids replicate to a much higher copy number. pUC vectors express the N-terminal fragment of the lacZ gene product and display α-complementation in appropriate hosts and recombinants can be identified by histochemical screening.

Bacteriophage vectors are the λ-phages and M13 phages and several versions have been developed. Bacteriophage λ infect *E.coli* and subsequent lysis allows propagation of cloned DNA. There are two methods of infection, the lytic phase and the lysogenic phase. In the lytic cycle, the phage injects its linear DNA into the cell, where it is ligated into a circle which replicate to form many phage particles that are released from the cell by lysis and and cell death. In lysogeny, the DNA integrate into the host genome by site-specific recombination. λ expression vectors (eg. λgt11) will accumulate the fusion protein in plaques and an immunological screening is used to identify the clones producing the protein of interest, by an antibody to the protein.

Box 6.8. LacZ′

LacZ′ is a shortened derivative of lacZ, which produces the N-terminal α-peptide of β-galactosidase. It produces a fusion protein of the N-terminal sequence followed by the protein sequence of the insert. It is however essential that the host strain contains a mutant gene that expresses only the C-terminal of β-galactosidase.

Box 6.9. λ-ZAP

It is an insertion vector designed to facilitate DNA and RNA sequencing and synthesis of RNA probes utilizing bacteriophage T7 and T3 promoters flanking the cloned fragments. DNA fragments up to 10 kb in length can be inserted in unique restriction sites. The MCS is located in the N-terminal portion of lacZ′, leading to a lac phenotype on IPTG/X-gal and produces a fusion protein for detection by antibody screening.

Box 6.10. λ-DASH

It contains a MCS and is used to clone 9 kb-22 kb DNA fragments. The polycloning site sequences are inverted with respect to one another. Inserts that destroy the Xho I or Bam HI site can be excised at flanking sites. The bacteriophage T3 and T7 promoters adjacent to the polycloning sites allow RNA probes to be made without subcloning.

pGEM vectors are transcriptional vectors that allow the in vitro transcription of a cloned fragment with T7 and SP6 phage promoters flanking an MCS and are recognized only by their corresponding bacteriophage RNA polymerases either of which may be used in vitro to transcribe the desired strand of the inserted fragment.

M13 phage vectors contain a 6.7 kb circular single strand of DNA. After infection of *E.coli*, the complementary strand is synthesized and the DNA is replicated as a double stranded circle, the replicative form with about 100 copies per cell. The cells continue to grow slowly and the ssDNA are continuously packaged and released from the cells as new phage particles. The same ss of the complementary pair is always present in the phage particle.

Cosmid vectors are hybrids of 2 *cos* sites from λ phage and about 5 kb of plasmid DNA and can accept 45 kb of insert DNA. The recombinant molecule is mixed with a λ packaging extract, a mixture of phage proteins which packages naked DNA into phage heads *in vitro*, producing infectious virions. *Cos* sites separated by 40-50 kb insert will be packaged. Less than 38 kb and more than 52 kb will not produce an infective λ virus particle. The recombinant λ is used to infect *E.coli* host cells. Inside the host, the cosmid circularizes and replcates as a large plasmid.

Phagemid vectors are a hybrid of the filamentous phage M13 and plasmids to produce a vector that can grow as a plasmid, and also be packaged as single stranded DNA in viral particles. Phagemids contain an origin of replication (*ori*) for double stranded replication, as well as an f1 *ori* to enable single stranded replication and packaging into phage particles. Many commonly used plasmids contain an f1 *ori* and are thus phagemids.

Box 6.11. *Cos* Sites

Cos sites are cohesive sites present at the end of the 48.5 kb λ genome and consist of 12 bp cohesive ends. They are asymmetric, but in other respects are equivalent to very large (16 bp) restriction sites. The *cos* ends allow the DNA to be circularized in the cell and can transform into a host cell, 1000 times more efficient than plasmid transformation.

Box 6.12. pBluescript Vector

pBluescript II is a phagemid multifunctional cloning vector available from Stratagene. It is designed to simplify and expedite gene cloning and analysis. It consists of an MCS sequence, pUC origin of replication, f1 phage origin of replication for ssDNA, lac promoter, lacZ′ (α-complementing portion of β-galactosidase) for blue/white screening, amp^r gene for insertional inactivation. The Kpn I and Sac I restriction sites allow directional cloning in either orientation and (+)/(-) is designated for the f1 orientation.

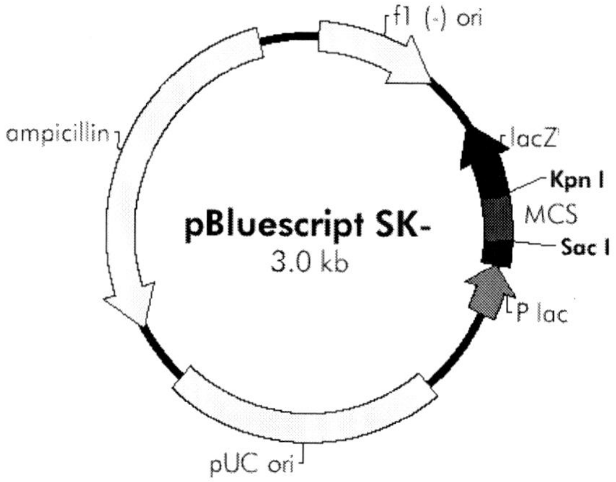

Box 6.13. BAC Vectors

Bacterial artificial chromosomes are based on the F factor of *E.coli* and can clone up to 350 kb of genomic DNA.

Box 6.14. YAC Vectors

Yeast artificial chromosomes can clone more than 1 Mb of target DNA. They contain two telomeric sequences (TEL), one centromere (CEN), one autonomously replicating sequence (ARS) and marker genes for selection in yeast.

Yeast episomal plasmid vectors are based on the 2 μ plasmid, corresponding to 6 kb of sequence length. These contain usually a 2 μ origin of replication, *E.coli* shuttle sequences and a yeast gene as a selectable marker such as the LEU2 gene, for replication in a yeast mutant which is auxotrophic for leucine. They normally replicate as plasmids or may integrate into a yeast chromosome by homologous recombination with the defective genomic copy of the selection gene.

Shuttle vectors can shuttle between a bacterial host as well as in an eukaryotic host. They incorporate the sequences required for replication and selection in *E.coli* or in yeasts. For example, pARC 145G contains a TRP1 gene for selection in yeast and an Amp^r gene for selection in *E.coli*. pARC 1520 contains URA3 gene for selection in yeast and an Amp^r gene for selection in *E.coli*.

Box 6.15. Yeast Expression Vector pESC

pESC yeast vectors allow cloning and expression of eukaryotic genes in yeast host strains such as *Saccharamyces cerevisiae*. pESC vectors are available from Stratagene which also provides the *S.cerevisiae* host strain. These are yeast episomal plasmids and use pBluescript II SK (+) as the phagemid backbone.

pESC-His consists of an MCS, yeast GAL1/GAL10 promoters, yeast HIS3 selection marker ORF, fl origin for ssDNA replication, yeast ADH1 and CYC1 terminators, pUC origin of replication and 2μ yeast origin of replication, bla ORF and tagged epitopes.

S.cerevisiae is approved by US Food and Drug administration as an GRAS (generally regarded as safe) organism.

Table 6.1. Compatibility of Different Vectors

Vector	Origin	Genome size	Insert size
Plasmid	Bacteria	10 kb	10 kb
λ phage	Viral	50 kb	23 kb
Cosmid	λ phage and bacteria	4–6 kb	45 kb
YAC	Yeast	11–12 kb	1000 kb

Table 6.2. Commercially available vectors

Vectors	Function
λ Max1, λ GEM2	Directional cloning
ZAP, Ziplox	Subcloning of cDNA inserts
λ gt10	Vector selection using DNA probes
λ gt11	Vector selection by immunoscreening
λ ZAPII	Blue/white phenotype

6.3. Techniques in Gene Cloning

6.3.1. Isolation of Plasmid DNA

Alkaline lysis is the commonly used method for isolation of plasmid DNA from fresh bacterial cells in their log phase of culture growth. It is advisable to follow an amplification procedure prior to this miniprep method to get sufficient amounts of plasmid DNA. A fresh inoculum of bacteria, *E.coli* is inoculated into LB broth and incubated O/N. In the log phase of this culture, it forms the inoculum for the next batch and can follow up to 4 inoculations which would amplify the number of cells in the final culture. Amplification kits are in use for rapid amplification and this would be convenient prior to the actual isolation method. Upon harvesting of bacteria, the cell pellet is resuspended in a buffer containing lysozyme that would digest the bacterial cell wall. This is then treated with a fresh alkaline solution containing SDS and NaOH. The preparation is neutralized with a concentrated potassium acetate solution at pH 5.0 and centrifuged and the lysate now contains the plasmid DNA. The plasmid DNA is concentrated further by addition of large volumes of ethanol (2-3 vols.) and air-dried.

Box 6.16. Plasmid DNA Isolation Reagents

Lysozyme	to lyse the bacterial cell wall
SDS	disrupts cell membrane and denatures proteins
NaOH	denatures linear chromosomal DNA, hydrolyse RNA
Potassium acetate	precipitate denatured proteins, chromosomal DNA and SDS
Ethanol	precipitate and concentrate DNA
70% ethanol	removes excess salts from DNA

The resulting pellet is now suspended in a Tris-EDTA buffer until further use. RNase A may also be added to the DNA preparation to digest any RNA contamination.

Box 6.17. TE Buffer

TE buffer is 10 mM Tris and 1 mM EDTA. When DNA is suspended in TE buffer EDTA chelates Mg2+ ions in restriction digestion. For procedures where Mg2+ ions are necessary, it is safe to suspend DNA in nuclease-free or ultrapure water.

Once plasmid DNA is isolated, it is essential to assess the purity of a DNA preparation and also arrive at its concentration. Pure DNA has an absorption maximum at 260 nm. Any contaminating protein is measured at 280 nm. A universal formula for predicting the purity of DNA is that its A_{260}/A_{280} should be 1.8-1.9. Lower ratios indicate protein contamination. The concentration of DNA is estimated based on the concept that an absorbance at 260 nm with an OD of 1.0 is equivalent to 50 µg of ds DNA /ml or approx. 37 µg of ssDNA/ml.

Spectrophotometric determination of DNA requires a concentration of 5 µg of DNA per ml. Absorbance of DNA at 260 nm is due to the chemical nature of the base moieties. Deoxyribose and phosphate groups do not absorb at A_{260}. Absorbance at 325 nm is due to particulates in DNA sample or dirty cuvettes, while absorbance at 230 nm is due to aromatic moieties such as phenol and small molecules containing peptide bonds and urea. Bio-Rad offers a SmartSpec Plus spectrophotometer that automatically reports the purity and concentration of a DNA sample placed in the cuvette. This is a convenient equipment and is recommended for use in a laboratory with routine DNA analysis.

Once the DNA sample is ascertained of its purity, it is necessary to analyse the DNA for its number of base pairs (kb) and its helical structure. Agarose gel electrophoresis of DNA is a very important tool and is a routine investigation in any molecular biology lab. This is similar to gel electrophoresis of proteins but uses agarose as the polymer matrix and forms a range of pore sizes suited for various DNA samples.

Table 6.3. Agarose Concentrations for DNA Fractionation

Agarose concentration (%)	Fractionation size (bp)	DNA sample
0.5	1000-30,000	Undigested genomic DNA
0.8	800-12,000	Digested genomic DNA
1.0	500-10,000	Plasmid DNA
1.2	400-7000	Plasmid DNA
1.5	200-3000	PCR products, plasmids
2.0	100-1500	PCR products

DNA is negatively charged owing to its acidic phosphate groups. In an electric current, DNA moves towards the anode. This is accomplished in a gel tank containing an appropriate buffer as the running buffer with the gel cassette. The gel cassette is previously cast with the required percentage of pure agarose, previously melted and cooled to 60°C, with addition of ethidium bromide which serves to visualize the DNA in UV light. Prior to cooling of the gel, a comb is placed for formation of wells which serve as the DNA sample reservoir.

Box 6.18. Ethidium Bromide (EtBr)

Ethidium bromide intercalates between DNA/RNA base pairs and in UV light emits light in the visible range and DNA is visualized. EtBr is a widely used stain and is mutagenic and requires necessary precautions while handling. EtBr can detect as little as 10 ng DNA. A stock of 5 mg/ml EtBr is prepared in water. A 10 μl of the stock can be added to 100 ml of agarose solution.

The gel cassette is immersed in the gel tank facing towards the anode (red) and the DNA samples are pipetted into the wells along with a DNA marker (ladder or digests). In the electric current, DNA migrates towards the anode and run as lanes in the agarose matrix. DNA samples and the marker are treated with a loading dye prior to sample loading to enable termination of the gel run visually. Usually, the electrophoresis is terminated half across the gel matrix and bands are visualized in a UV transilluminator. If DNA bands are observed the run is stopped or otherwise, allowed to continue the run for a longer time period.

The DNA bands are documented in a gel documentation system to determine the DNA size, percentage of DNA bp in each band, total amount of DNA in each band and quantity of unknown DNA by comparison with intensity of DNA marker.

Box 6.19. Loading Dye

Loading dye makes the DNA samples denser than the running buffer and makes it sink to the bottom of the well and for visualization of the progress of the electrophoresis run. Fragments of DNA approx. 2,800 bp migrate with the bromophenol blue and fragments of approx. 250 bp migrate with xylene cyanol. A 10X loading dye is prepared and an aliquot is added to samples.

Box 6.20. Acridine Orange

Acridine orange is a fluorescent stain that is used to determine whether a DNA sample is double stranded or single stranded. Addition of acridine orange to DNA in a UV light produces a green fluorescence for dsDNA or an orange fluorescence for a ssDNA in an agarose gel electrophoresis.

2% acridine orange , about 4 μl is added to 1 ml of DNA sample.

Acridine orange is an intercalating agent that alters the reading frame of DNA (mutagenic) and hence, samples may not be recovered.

Plasmid DNA exists inside the bacterium in supercoiled configuration. Upon isolation, it may be supercoiled, relaxed circular when nicked (with DNase) or linear. Supercoiled plasmid travels fastest and farthest, relaxed circular travels slowest and linear DNA travels midpoint reflecting size in an agarose electrophoretic gel.

6.3.2. Restriction Enzyme Digestion

A restriction digest is performed to produce DNA of different sizes and yields a restriction pattern with the use of restriction enzymes at appropriate conditions in an appropriate buffer. Generally, restriction enzymes are supplied with diluent buffers and an assay buffer. Restriction digests may be with single restriction enzyme or two different restriction enzymes. The restriction digests indicate the sequence of DNA patterns, mapping of vectors, excise target DNA for cloning and splitting of large DNA into DNA fragments. A restriction digest consists of about 1 μg of DNA and 20-25 units of restriction enzyme(s). Some restriction enzymes require BSA or acetylated BSA for the reaction and this can be referred with the supplier catalogs. After incubation, it is necessary to stop the reaction by heat-inactivation, EDTA or by phenol/chloroform extraction. Digested DNA may be concentrated by ethanol precipitation and electrophoresed in a 1% agarose gel.

One unit of the restriction enzyme is defined as the amount of enzyme that can digest 1 μg of phage λ DNA to completion in 1 h in a 50 μl reaction, using the specified buffer and at the specified temperature.

Restriction enzymes should be stored at -20°C or below. Restriction endonucleases require Mg^{2+} ions for catalysis but no other cofactors or coenzymes and an optimal pH of 7.2-7.8. Typical digests require 1-2 h incubations.

6.3.3. Ligation

Ligation is the covalent joining of a DNA fragment to a vector with compatible ends which may be blunt ended (blunt ligation) or 5'-overhangs (cohesive end ligation) by a DNA ligase. Ligation requires a free 3'-OH on one DNA fragment and a 5'-phosphate on the other to form a covalent phosphodiester bond between the two DNA fragments using T4 DNA ligase. Blunt end ligation requires that the same restriction enzyme need not be used to generate the DNA fragments. Blunt-ended fragments associate by random collisions and a restriction enzyme site is not regenerated. Cohesive end ligation involves complementary base pairing of vector DNA and insert DNA overhangs and the same restriction enzyme is required to cleave the insert and the vector DNA to generate complementary or sticky ends and the same restriction site is regenerated after ligation. DNA ligase requires an adenylating agent to activate the phosphate group for attack by 3'-OH. *E.coli* ligase requires NAD^+ while, T4 phage ligase is ATP-dependent. During ligation, it is probable that vector ends may be religated and recircularize, instead of ligation to the DNA insert. To prevent this, vector dephosphorylation is necessary prior to a ligation protocol. This is achieved by an alkaline phosphatase that removes phosphate groups from the 5'-ends of vector DNA (see Section I, Experiment VIII). The ligated vector is now ready for transformation in a suitable host.

6.3.4. Transformation

Transformation is a process where cells uptake recombinant plasmids for expression of the proteins. Cells are made competent to facilitate the transformation. The transformation step is a very critical process and has to be carried out in specific timings and with appropriate guidelines. The number of transformants indicates the efficiency of transformation in an LB plate with the transformed cells appropriately diluted. Transformation procedures have been developed for bacterial as well as yeast host cells. Transformation in bacteria is simpler with ice-cold $CaCl_2$ induced competent cells of *E.coli* and uptake of plasmids by heat shock, while yeast cells require an agent to specifically penetrate the cell wall such as lithium acetate and PEG due to complexity of its cell wall structure consisting of polysaccharides and chitin, and uptake by heat shock. Heat shock at 42^oC enables rapid uptake of plasmids by the competent cells and the timing is critical. An important consideration of transformation is that upon heat shock, the transformed cells should be allowed to stand in LB at RT for 30-45 min for subsequent repair of the host cell wall by the action of DNA ligases. The cells are now ready for plating in LB agar and transformation efficiency is determined after growth of the transformants. Transformation experiments require an experimental plate, a negative control plate and two positive control plates. The following step is the identification of recombinant plasmids by a procedure known as blue-white screening which indicates recombinant white colonies against non-recombinant blue colonies in the presence of a substrate X-gal and an inducer IPTG in the LB agar plate medium and picked for colonies by replica plating. Another screening method to identify recombinants is by enzyme plate assay. For example, for lipases tributyrin-agar or Rhodamine-tributyrin agar can be followed.

Transformation in Bacteria

Bacterial cells are made competent by incubation with ice-cold $CaCl_2$. This apparently makes the cell envelope permeable. Cold $CaCl_2$ is important for bacterial cell competence. Transformation is achieved by heat shock at 42^oC and incubation at RT. Heat shock alters the cell envelope, allowing the plasmid to penetrate inside, then closes the cell envelope again. A short period of incubation in LB enables the cells to recover from the experimental manipulations. Another method of transformation is electroporation which uses a high voltage charge to make transient holes in the cell membrane of bacteria through which the plasmid uptake takes place. The procedure for preparation of competent cells is given below as adapted from Methods in Enzymology, Vol. 152 (1987):

Grow a fresh O/N culture of cells in LB broth at 37°C. Dilute the cells 40-fold into 1L fresh medium. Incubate at 37°C with good aeration until an absorbance at 550 nm of 0.4-0.5 is reached. Immediately chill the culture by swirling in an ice-water bath. When the cells are chilled, centrifuge the culture at 4°C at 5000 rpm for 10 min. Decant the supernatant and place the pellets on ice. Resuspend the pellets in 500 ml of ice-cold 100 mM $CaCl_2$ in 0.1 M Tris.HCl buffer, pH 7.4. It is easier to resuspend the pellets if they are first vortexed before the $CaCl_2$ is added as some clumping can occur. The cells can be uniformly suspended by sucking up and down with a 25 ml pipette. Once the cells are resuspended, incubate in ice for 30 min with occasional swirling. Pellet the cells once again at 5000 rpm for 10 min at 4°C. Resuspend in 40 ml of ice-cold 100 mM $CaCl_2$, 15% glycerol. Distribute aliquots of 0.2 ml cells into sterile eppendorf tubes in ice. Keep in ice at 0-4°C for 12-24 h. This step is essential for high competency although the cells are considerably competent at this stage. Freeze the tubes in ethanol-dry ice or liquid nitrogen and place immediately at -70°C. The cells retain their competency for months if kept in this manner.

Transformation in Yeast

Saccharomyces cerevisiae is an ideal host for expression of eukaryotic proteins and has a generation time of 90 min and can grow on liquid/semisolid media in rich medium and at lower temperatures around 25-30°C. *Saccharomyces cerevisiae* is approved by the US Food and Drug Administration as a GRAS organism (generally regarded as safe).

Transformation in yeasts may be by cell competency, electroporation and spheroplasts. Cell competency requires lithium acetate, an alkali metal cation for the yeast cells to become competent, single stranded carrier DNA and PEG to introduce the plasmid DNA. Carrier DNA increases transformation efficiency by one or two orders of magnitude. PEG treatment coupled with heat shock makes the cell wall permeable and triggers DNA uptake. Electroporation is extremely efficient, especially when small quantities of DNA are available. Spheroplast transformation involves removal of cell wall by enzymatic treatment. DNA is introduced with the resulting spheroplast with PEG. Spheroplasts must be handled gently and can be easily disrupted by vortexing, vigorous pipetting and osmotic shock. Transformed spheroplasts are plated on TOP AGAR, a thin layer of softer (lower concentration) agar on top of the regular agar layer in plate, to provide an environment for regeneration of the cell wall. Spheroplast transformation provides greater transformation efficiency. Identification of transformed yeasts by use of auxotrophic yeasts is efficient with Trp^- and Ura^- markers. The auxotrophic yeast mutants require Trp or Ura supplements for growth.

Transformation efficiency is defined as the number of transformants per μg input plasmid DNA per 10^8 cells. Transformation efficiency is determined by calculating the number of transformants in 1.0 ml of resuspended cells per 1.0 μg plasmid per 10^8 cells.

Box 6.21. Transformation in Yeast

Grow 100 ml cells to A_{600}=0.5-0.7 (A_{600} 0.7=2x10^7 cells/ml). Pellet at 3000 rpm for 5 min. Wash with LiAc mix and pellet. Resuspend in 1 ml LiAc mix. Transform using 100 µl cells per transformation. Add 1 µg DNA in <15 µl and 50 µg carrier DNA. Add 0.7 ml PEG mix and mix by vortexing. Incubate at 30°C for 30 min. Heat shock at 42°C for 15-20 min. Pellet in microfuge and resuspend with 0.2 ml microtip to disperse pellet in 200 µl water. Plate in LB agar for transformants to grow O/N.

LiAc mix: 100 ml lithium acetate in TE buffer (1.02g/100 ml)

PEG mix: 40% PEG in 100 mM LiAc/TE (4g of PEG in 7 ml TE buffer (prepared freshly)

Carrier DNA: Calf thymus DNA (autovclaved, boiled fresh before use 4 min earlier)

Box 6.22. Calculation of Transformation Efficiency

1.0 x 10^8 cells with 100 ng of plasmid yields 500 colonies on yeast dropout medium spread with 1 µl of cell suspension dispersed in 100 µl of sterile water.

Transformation efficiency = 500 x 1000 (plating factor) x 10 (plasmid factor) x 1 (cells/transformation x 10^8) = 5 x 10^6 transformants /1.0 µg plasmid /10^8 cells.

6.4. PCR AND RT-PCR

6.4.1. Primers

Primers are short sequences of DNA that anneal to single strands of DNA and extend along the DNA by adding dNTPs in the presence of a DNA polymerase. This forms the basic principle of PCR where the newly synthesized DNA strands act as another template for primer molecules with continuous synthesis of DNA and thereby amplify the DNA several fold. Primers are of significance in molecular biology especially in PCR and RT-PCR. Primers are of 17-30 base sequences known as 30-mer or may be 30-45 base sequences known as oligonucleotide probes. DNA probes are highly used in hybridization techniques and southern blots.

For PCR, primer pairs have to fulfil certain criteria that are essential for efficient amplification of DNA and prevent formation of mismatched base addition. A hairpin loop structure in the primer pair should be avoided for self-complementary base pairing. Complementary sequences at 3'-ends of the primers result in primer dimers. Following certain rules for primer design enables a successful PCR amplification and obtain specific PCR products. Hence, we describe some facts to be considered during primer design for PCR or RT-PCR:

- Primers should be 17–28 bases in length
- Base composition should be 50–60% (G+C)
- 3′-ends of the primer should be G or C or CG or GC which prevents 'breathing' of ends and increases efficiency of priming

- Melting temperatures of primers should be Tm=55–80°C
- Presence of 3 or more G or C at 3′-ends promote mispriming at G or C-rich template sequences
- 3′-ends of primers should not be complementary which yield primer dimers
- Primer self complementarity should be avoided which tend to form a hairpin loop structure.

Primer design softwares are innumerous available from the internet. GCG is an analysis package that determines the efficiency of a primer designed for a suitable portfolio and predicts any hairpin loop that form secondary structures in the primer pair.

The melting temperature of primers is a critical criterion that determines the annealing temperature of the PCR reaction. The annealing temperature of the primer should be 5°C below the lowest Tm of the pair of primers. A general formula for determination of the Tm of <18-mer primers is,

$$Tm = 4(G+C) + 2(A+T)°C$$

and for >18-mers is,

$$Tm = 64.9°C + 41°C \times (G+C- 16.4)/N,$$ where N is the length of the primer.

Primers may also be analysed for their Tm using calculators available in the internet such as the Promega primer calculator available at http://www.promega.com/biomath/calc11.htm and the Finnzymes Tm calculator at https://www.finnzymes.fi/tm_determination.html In PCR, primers are used at pmol concentrations and it is necessary to convert the primer concentration of ng concentration to pmol concentration. The formula to obtain 50 pmol primer available from Promega is,

50 pmol=16.3 ng x b, where b is the number of bases in the primer.

6.4.2. *Taq* Polymerase

DNA polymerase is a functional component in PCR that actively adds a dNTP to the extending DNA strand in a 5′→ 3′ direction along the template strand. Earlier polymerases were from *E.coli* which is unstable notwithstanding the denaturation temperatures employed. It was identified in 1965 by Thomas D. Brook that a highly thermostable bacterium, *Thermus aquaticus* from hot springs produced a thermostable DNA polymerase and is known as the *Taq* polymerase. This relatively simplified the PCR conditions and is largely used in many of the laboratories. The temperature optimum for polymerase activity is 75–80°C, with a half-life of 9 min at 97.5°C, and can replicate a 1000 base pair strand of DNA in less than 10 seconds at 72°C. One of its drawbacks is its relatively low replication fidelity. It lacks a 3' to 5' exonuclease proofreading activity of the Klenow fragment of the polymerase. Thermostable DNA polymerases have been isolated from other thermophilic bacteria and archaea, such as *Pfu* DNA polymerase, possessing a proofreading activity, and are being used instead of or in combination with *Taq* for high-fidelity amplification.

6.4.3. PCR Cycle

PCR stands for 'Polymerase Chain Reaction' used for amplification of DNA fragments by consecutive steps or cycles. First developed by Kary Mullis in the 1980s, it has been made automated and is now a routinely used laboratory protocol. The basic PCR consists of repeated cycles of denaturation, annealing, polymerization and a final extension to complete elongation. Denaturation is the step of separation of the double helical DNA strand to allow primers to bind to complementary sequences in the template. An initial denaturation of 95°C for 1 min is recommended followed by another denaturation step for 30s at 94°C in cycles. After denaturation, annealing of the primers to the complementary sequences is effected usually at 55°C depending upon the Tm of the primers. The next step is the polymerization or elongation where addition of dNTPs to the extending primer takes place in a 5′→ 3′ direction by the polymerase activity at generally 72°C for 1 min. This sequence of denaturation (94°C, 30s), annealing (55°C, 30s) and polymerization (72°C, 1 min) is one PCR cycle and 25–30 cycles of these steps allow rapid amplification of the target DNA to yield sufficient PCR products. A final elongation step at 72°C for 3–5 min is required to ensure completion of the strand synthesis. Thus, PCR can amplify DNA fragments to the order of 10^9.

6.4.4. Factors Affecting PCR

A range of factors have to be considered prior to a PCR procedure that makes a successful PCR protocol. PCR optimization is a requisite step to achieve full length PCR products and increases the efficiency of a PCR reaction. A revision of the factors/conditions that affect PCR is therefore essential.

Denaturing Temperature and Time

Annealing of primers to the template DNA is the critical property of a PCR and depends upon the denaturation temperature (Tm) of the primer pairs. Buffers of low ionic strengths such as in 150 mM NaCl, the Tm is generally less than 100°C, and hence, PCR works at denaturation temperatures of 91–97°C. Time at temperature is a factor that augments the efficiency of PCR. By keeping a lower time limit for denaturation at 94°C, one may increase the number of amplification cycles. This is especially suited for short templates where denaturation time may be reduced to 30 s or less.

Annealing Temperature and Primer Design

The annealing temperature depends upon the primer length and G+C content. It is advisable to design primers with similar G+C content so that there will not be a wide gap between the Tm of the individual primers. The annealing temperature should be 5°C below the lowest Tm of the primer pair.

Primer Length

The optimum length of a primer depends upon its A+T content and the Tm of the primers. The primers should be complex enough so that the annealing to sequences other than the target sequence is very low. A sixteen base sequence will statistically be present only once in every 4^{16} bases ($4x10^9$). Therefore the association of a greater than 17-mer with its target sequence is extremely sequence specific.

Elongation Temperature and Time

Elongation temperature and time is usually 70–72°C for 0.5–3 min, with primer extensions of upto 100 bases per second. Longer sequences require longer times with about 3 min for 3 kb sequences.

Cycle Number

The number of amplification cycles (for a band to be visible in an agarose gel) depends on the starting concentration of template DNA. About $3x10^5$ molecules require 25–30 cycles and about 50 molecules require 40–45 amplification cycles.

Helix Destabilizers/Additives

Template DNA with high G+C content requires harsh denaturing conditions. A 10% DMSO/DMF/urea /formamide may be used in the reaction mix which lowers the Tm of the target DNA. However, DMSO is known to decrease the polymerase activity of *Taq* by 50%.

Box 6.23. Additives in PCR mix

DMSO	amplification of >1kb products
Formamide	improve the specificity of PCR
Glycerol	amplification of high G+C templates
PEG	promotes macromolecular association at low template concentration

Optimization

PCR reactions are not usually 100% efficient and optimization of reaction conditions is necessary. Annealing temperature and Mg^{2+} concentration are the main factors to achieve proper optimization. Too low annealing temperatures favour mispairing. Free Mg^{2+} concentration should be 0.5–2.5 mM > dNTPs. PCR optimization kits are available which simplifies elaborate running of the PCR and increase the specificity of the actual PCR protocol for a full-length PCR product.

6.4.5. PCR Protocol

A basic PCR protocol consists of the template DNA, a primer set, polymerase, reaction buffer for optimum activity of the polymerase, dNTPs, $MgCl_2$ and nuclease-free water. Addition of the components in an order and in an ice-bath is recommended. A common PCR protocol by Promega is given in Table 6.4:

Table 6.4. Component addition for PCR

Components	Volume to be added
10X reaction buffer	5 μl
dNTP mix (10 mM of each dNTP)	1μl
Taq DNA polymerase (5 u/μl)	0.25 μl
25 mM $MgCl_2$	3 μl
Forward primer	50 pmol
Reverse primer	50 pmol
Template	<10 ng/ μl
Nuclease-free water	Final volume of 50 μl

PCR is carried out by denaturation, annealing and extension at appropriate time and temperature cycles. The PCR products are analysed by an agarose gel prior to which a PCR purification kit may be employed to eliminate unreacted components and elute only the PCR product.

6.4.6. Reverse Transcription

The Central Dogma states that the flow of genetic information follows from DNA to RNA (transcription) and RNA to proteins (translation) postulated by Francis Crick. Later, it was identified that retroviruses possessed the cellular machinery for reverse transcription, i.e., RNA was reverse transcribed to DNA by the viral enzyme reverse transcriptase. The resulting DNA is termed complementary DNA or simply cDNA. Reverse transcriptase is an RNA-dependent DNA polymerase which was discovered independently by David Baltimore and by Howard Temin in 1970. It also helps in the formation of a double helical DNA once the RNA is reverse transcribed into a single strand cDNA. This discovery led to the innumerous applications of reverse transcriptase, and one such application is RT-PCR where RNA isolated from cells or tissues forms the template for cDNA synthesis using Moloney Murine Leukemia virus (M-MuLV) or Avian Myeloblastis virus (AMV) reverse transcriptase under suitable conditions, in the presence of a sequence-specific primer. The RNA can then be degraded by RNase H that actively cleaves the RNA from RNA-cDNA hybrid. Total RNA can also be used for reverse transcription using an oligo$(dT)_{18\text{-}23}$ primer that binds effectively to the mRNA species containing 3′-poly (A) tails. PCR can then be followed from the cDNA template for further amplification by a DNA polymerase. Recent protocols use random hexamer primers that bind to RNA sequence in a random fashion and reverse transcribe to a full-length cDNA. Also, several kits have been developed using the two reverse transcriptases for RT-PCR. An additional component is an RNase inhibitor that prevents degradation of the template RNA by RNases for a successful reverse transcription process.

Box 6.24. M-MuLV Reverse Transcriptase

M-MuLV enzyme possesses a RNase H activity specific to RNA in RNA-DNA hybrid. It is used in first strand cDNA synthesis, DNA labelling, analysis of RNA by primer extension.

One unit of the enzyme incorporates 1 nmol of dTMP into a polynucleotide fraction (adsorbed on DE-81) in 10 min at 37°C.

The storage buffer contains 50 mM Tris.HCl (pH 8.3), 0.1 M NaCl, 1 mM EDTA, 5 mM DTT, 0.1% TritonX-100 and 50% glycerol. The 5X reaction buffer contains 250 mM Tris.HCl (pH 8.3) at 25°C, 250 mM KCl, 20 mM $MgCl_2$, 50 mM DTT. It is inactivated by heating at 70°C for 10 min. It is inhibited by phosphate, pyrophosphate and polyamines.

Box 6.25. RNase H

RNase H specifically degrades only the RNA strand in a RNA-DNA hybrid. It does not hydrolyse the phosphodiester bonds within ss- and ds-DNA and RNA. It is used in the removal of mRNA prior to the synthesis of the second strand of cDNA, removal of poly (A) sequences of mRNA after hybridization with oligo (dT), site-specific cleavage of RNA, studies of in vitro polyadenylation reaction products.

One unit of the enzyme catalyses the formation of 1 nmol of acid soluble products in 20 min at 37°C.

The storage buffer is 25 mM HEPES-KOH (pH 8.0), 50 mM KCl, 1 mM EDTA, 1 mM DTT, 0.1 mg/ml BSA and 50% glycerol. It is inactivated by heating at 65°C for 10 min.

6.4.7. RT-PCR Protocol

This procedure is similar to the PCR protocol, except that reverse transcription of the synthesis of cDNA precedes the usual PCR protocol. RT-PCR can also be carried out with one sequence-specific primer in the presence of an oligo (dT) primer. Additional components other than the kit components could include RNase H and RNase inhibitor. Finnzymes RT-PCR kit protocol is given in Table 6.5:

Table 6.5. Component Addition for RT-PCR

Components	Volume to be added
10X RobusT reaction buffer	5 µl
50 mM $MgCl_2$	1.5 µl
dNTP mix (10 mM of each dNTP)	1 µl
RNase inhibitor	x µl
Template RNA	10 pg-1 µg
Forward primer	10 pmol
Reverse primer	10 pmol
M-MuLV RT (RNase H) 5 U/µl	2 µl
DyNAzyme EXT 1 U/ µl	2 µl
Nuclease-free water	Final volume of 50 µl

cDNA synthesis is carried out at 37–50°C for 1 h followed by denaturation, annealing and extension at appropriate time and temperature cycles. The RT-PCR products are analysed by an agarose gel prior to which a purification kit may be employed to eliminate unreacted components and elute only the PCR product.

6.4.8. PCR Methods

The conventional PCR has been modified to suit various applications of research, analytical and commercial interests:

Nested PCR

PCR reactions may produce mispriming due to sequence similarities between the target DNA and related sequences in primers (with a low annealing temperature) yielding smears in an agarose gel. This can be overcome by a nested PCR, where a second set of primers is used in an aliquot of the original PCR mix, which have sequences 3′ to the first set and anneal within the previously amplified fragment giving a shorter fragment but without the contaminating sequences.

Hotstart PCR

Taq polymerase possess a degree of activity at RT which can lead to imperfectly matched duplexes prior to the actual PCR process. This artefact can be set right by the addition of a critical component after the denaturation step in a PCR reaction. This kind of PCR is known as Hotstart PCR using the enzyme or Mg as the critical component. Cross contamination while opening the thermal cycler may be avoided by using an inert wax to cover the reaction mixture and addition of the critical component on the top of the wax which melts during denaturation, and thermal convection mixes the reagents and PCR begins.

Inverse PCR

This method of PCR can be used to amplify sequences that are upstream or downstream of the region amplified by the basic primer pair.

Multiplex PCR

This PCR variation uses multiple pairs of primers, to amplify more than one DNA target sequence in the same reaction and the resulting PCR fragments are distinguished in an agarose gel by PCR products of different lengths.

Quantitative PCR

This type of PCR is used to determine the number of DNA molecules in a test sample.

PCR Mutagenesis

PCR can be used to introduce specific mutations into a DNA template by using primers differing in base composition, i.e., non-similar base sequences to the template.

6.5. TECHNICAL TIPS

- A_{260}/A_{280} of 1.8–1.9 indicates very pure DNA and 1.9–2.0 indicates very pure RNA.
- All restriction enzymes and nucleases require Mg^{2+} ions.
- An internal restriction site would be advantageous for design of primers.
- Ribonuclease A does not require Mg^{2+} ions.
- *Taq* polymerase does not have the 3′→5′ exonuclease or proof reading activity of Klenow fragment.
- Longer the target sequence between primers, the less efficient the PCR.
- Glycerol and DMSO reduce hairpin loops in secondary structure but can inhibit polymerases.
- BSA can enhance PCR by protecting the polymerase and chelating inhibitors.
- Pancreas contains high level of RNases.
- The abundance of mRNA determines the size of cDNA library.
- The majority of mRNA species are between 1.5–2.0 kb in length.
- Directional cloning is necessary for expression cloning.
- Phenol and SDS inhibit cDNA synthesis.
- It is essential to add RNase inhibitor RNasin, to improve the length of first strand transcripts when using AMV reverse transcriptase, but it is optional for M-MuLV reverse transcriptase.
- First strand synthesis reaction at 50°C reduces secondary structure in the RNA template which may cause the reverse transcriptase to halt.
- Higher transformation efficiency is in the order of 10^9 cfu/µg. Lower transformation efficiency is between 10^6–10^7 cfu/µg.
- Sau3AI is inhibited by β-mercaptoethanol.
- A simple reversible way to stop a restriction digest is to add EDTA. Each EDTA molecule binds to two Mg^{2+} ions and prevents catalysis.
- An irreversible method to inactivate restriction enzymes is by phenol extraction.
- Ethanol precipitation is used to remove residual phenol.
- Phenol, chloroform, alcohol, detergents, EDTA, inhibit restriction enzyme activity.
- *E.coli* strain MM294 is commonly used in academic and research labs.
- A simple glycerol stock is 2X glycerol storage buffer containing 25 mM Tris, pH 8.0, 0.1 M $MgSO_4$, 65% glycerol. 1 ml of GSB is added to 1 ml of culture and stored at -20°C.
- DNA fragments of approx. 2800 bp migrate with bromophenol blue and fragments of approx. 250 bp migrate with xylene cyanol.
- Warm water, about 50–60°C will activate yeast.
- NaOH-SDS solution used for plasmid minipreps should be prepared fresh.
- Ice-cold $CaCl_2$ is important for bacterial cell competence.
- Incubation of *E.coli* at 30°C prevents overgrowth.

6.6. A Cloning Strategy for Lipase by RT-PCR Cloning

It is a simple case for the cloning of lipase genes in a bacterial source, which calls for genomic DNA isolation, PCR, ligation and transformation. However, a lipase gene from an eukaryotic source presents complications due to the presence of introns or the intervening sequences and the resulting product may just be a noncoding region of the gene product. Hence, cloning procedures using cDNA which are products of mRNA that is spliced of introns during transcription may be employed. A method to produce a cDNA library and later screen the library for the desired gene (lipase) or by isolation of total RNA followed by RT-PCR would be the methods of choice for eukaryotic lipase cloning. We describe RT-PCR cloning method for an eukaryotic lipase source using FINNZYMES RT-PCR kit and Promega protocols. Transformation protocol is followed according to Short Protocols in Molecular Biology.

6.6.1. Calculation of Concentrations for RT-PCR Components

It is very important in molecular cloning that the final concentrations of the PCR/RT-PCR components are calculated prior to the start of the cloning strategy. This allows manipulations to be checked and referenced and enables optimization of the protocol in case of any errors during or after the procedure. We shall consider the calculations of FINNZYMES RT-PCR components:

10X RobusT reaction buffer
5 μl of 10X buffer added to 50 μl final reaction volume = 1X
50 mM $MgCl_2$
1.5 μl of 50 mM $MgCl_2$ added to 50 μl = 166.6 μM
dNTP mix (10 mM each)
1 μl of dNTP added to 50 μl= 200 μM
Template RNA
10 pg to 1 μg in 50 μl
Forward primer
10 pmol in 50 μl
Reverse primer
10 pmol in 50 μl
M-MuLV Reverse transcriptase (5 U/μl)
2 μl added to 50 μl = 10 U/50 μl or 0.2 U/μl
DyNAzyme EXT (1U/μl)
2 μl added to 50 μl = 2 U/50 μl or 0.04 U/μl
RNase inhibitor (40 U/μl)—optional
2 μl of (2.5 μl to 5 μl stock) to 50 μl = 20U/50 μl or 0.4 U/μl

Table 6.6. An RT-PCR Protocol for Lipase

Components	Volume to be added	Final concentration
10X RobusT reaction buffer	5 µl	1X
50 mM $MgCl_2$	1.5 µl	166.6 µM
dNTP mix (10 mM of each dNTP)	1 µl	200 µM
RNase inhibitor	2 µl	20 U
Template RNA	10 pg-1 µg	10 pg – 1 µg
Forward primer	10 pmol	10 pmol
Reverse primer	10 pmol	10 pmol
M-MuLV RT (RNase H) 5 U/µl	2 µl	10 U
DyNAzyme EXT 1 U/ µl	2 µl	2 U
Nuclease-free water	Final volume of 50 µl	

6.6.2. RT-PCR Protocol

The application protocol is followed by keeping all the components, reaction mixes and samples on ice. The following reaction components are added to nuclease-free microfuge tubes placed on ice:

Gently mix the components and spin briefly. Depending on the thermocycler used, overlay with mineral oil if required. A typical temperature profile is given below:

(1) cDNA synthesis, inactivation of M-MuLV reverse transcriptase and denaturation of the cDNA/RNA hybrid:
    1 cycle of:
    37–50°C for 15–60 min
    94°C for 2 min
(2) PCR amplification:
    25–40 cycles of:
    Denaturation: 94°C for 30 s
    Annealing: 30 s at Ta of primers
    Extension: 72°C for 1–1.5 min per kb
(3) Final extension
    72°C for 5–10 min

(Annealing temperature Ta is dependent on the melting temperature Tm of the primers used and the cycle number is dependent on the abundance of the target RNA).

Control Reactions

To monitor the RT-PCR and facilitate optimization, both positive and negative control reactions have to be performed.

Table 6.7. A Positive Control RT-PCR Protocol (Finnzymes)

Components	Volume to be added
10X RobusT reaction buffer	5 μl
50 mM $MgCl_2$	1.5 μl
dNTP mix (10 mM of each dNTP)	1 μl
RNase inhibitor	2 μl
Control RNA	1 μl
Forward control primer10 pmol	1 μl
Reverse control primer10 pmol	1 μl
M-MuLV RT (RNase H) 5 U/μl	2 μl
DyNAzyme EXT 1 U/ μl	2 μl
Nuclease-free water	Final volume of 50 μl

Positive Control Reaction

Control RNA and upstream and downstream control primers are supplied with the kit. The control RNA template is MS2 viral RNA with carrier (E.coli rRNA). A 1011 bp sequence is obtained in the positive control.

Negative Control Reaction

An identical reaction without M-MuLV reverse transcriptase is the negative control. A negative control is performed to exclude previously amplified products of a DNA contamination. No PCR product is observed.

Positive Control Protocol

(1) cDNA synthesis, inactivation of M-MuLV reverse transcriptase and denaturation of the RNA/DNA hybrid:
48°C for 30 min
94°C for 2 min

(2) PCR amplification:
25 cycles of 94°C for 30 s
60°C for 30 s
72°C for 1 min

(3) Final extension:
72°C for 7 min

An agarose gel for the RT-PCR product is run with MW markers to estimate its base pair size. The reaction mix is purified to remove all the RT-PCR components by a purification kit and the product is now ready for ligation with vector. Alternatively the product may be excised from the agarose gel.

6.6.3. Vector Digestion

A suitable vector such as pUC18 may be used as the cloning vehicle. Digestion of the vector with an appropriate restriction enzyme compatible to the insert DNA sequence is followed:

(1) Vector DNA 5 μl
10X restriction enzyme buffer 5 μl
Acetylated BSA (1μg/μl) 5 μl
Restriction enzyme (4-5 U/μg DNA) 20-25 U
Deionized water Final volume of 50 μl

(2) Incubate at appropriate temperature for 1–2 h

(3) 5 μl of the sample is run on a 1% agarose gel to check for complete digestion. For vector dephosphorylation (Section I, Experiment No. VIII), the restriction digest should be heat-inactivated and purified. The vector DNA may be concentrated by ethanol purification.

6.6.4. Ligation

The vector and the PCR product are ligated to form the clone by blunt-end or cohesive-end ligation. A simple ligation protocol from Promega is given below:

(1) A 1:1 insert:vector ratio is used. Typical ligations use 10–200 ng of vector DNA.

(2) The reaction mix consists of
Vector DNA (3 kb) 100 ng
Insert DNA (0.5 kb) 17 ng
T4 DNA ligase 10X buffer 1 μl
T4 DNA ligase (Weiss units) 1 U
Deionized water Final volume of 10 μl

(3) The ligation reaction is carried out at 4°C O/N or 15°C for 4–6 h or 20–22°C for 1–3 h.

Following ligation, the ligated vector DNA has to be transformed with previously prepared competent cells of a suitable host strain, *E.coli* BL21.

6.6.5. Transformation

Transformation of *E.coli* BL21 with the ligated vector is followed accordingly:

(1) Add 10 ng of vector DNA (10–25 μl) into 15 ml sterile, test tubes and place on ice. Rapidly thaw the competent cells by warming between hands and dispense 100 μl immediately into test tubes containing DNA, swirl and place on ice for 10 min.

(2) Heat shock cells by placing the tubes for 2 min in a 42°C water bath. Add 1 ml of LB broth to each tube. Incubate for 1 h at 37°C at 250 rpm.

(3) Several dilutions of transformed DNA are placed on appropriate antibiotic-containing plates and allowed to incubate for 12–16 h at 37°C. Calculate transformation efficiency.

Once transformed in the host strain, the cloned lipase gene is now expressed in a suitable expression system to elaborate lipase production that has been overexpressed. Lipase expression is optimized considering various factors that are derived from the host-vector compatibility. The expressed lipase gene may be purified and studied for its structure-function properties that encompass a vast area of investigation.

Self-evaluation

(1) Plasmids are
(a) extrachromosomal (b) capable of autonomous replication (c) both the above (d) none of the above

(2) Lac promoter is
(a) for blue-white screening (b) present in expression vectors (c) both the above (d) none of the above

(3) Acridine orange distinguishes
(a) ss DNA (b) ds DNA (c) both the above (d) none of the above

(4) Lysozyme can be used to
(a) lyse the bacterial wall (b) precipitate DNA (c) remove excess salts (d) all the above

(5) A common restriction enzyme from *E.coli* is
(a) Eco RI (b) Bam H I (c) Pst I (d) Hind III

Answers

1c 2c 3c 4a 5a

Chapter 7

BIOINFORMATICS

7.1. COMPUTING IN BIOINFORMATICS

7.1.1. Internet

The internet is a global network of computer networks that is used in effective communication worldwide with ease and precision. The internet application is based on TCP/IP (Transmission Control Protocol/Internet Protocol) shared communication protocols. This has enabled rapid and advanced progress in creation of various computer applications and one such is Bioinformatics. Many databases are freely available over the internet as handy tools for bioinformaticians and Life Sciences research community. Information has been made Public and is globally shared through computer networks by a World Wide Web or W3 developed at CERN, an European Laboratory for Particle Physics. The sharing of information is based on an IP address allocated to each computer terminal that identifies the user machine. The IP address is intelligible to the computer, however, an alternative user-friendly domain to each web address is unique to the site where it is located and information is retrieved by the end-user. Domain names are exclusive to all countries with a suffix to its web address, such as:

Australia uses the domain name	.au
Germany	.de
Switzerland	.ch
Japan	.jp

The domain names are also in common use for educational institutions as .edu, commercial industries as .com and organizations use .org. It is of significant interest that the web addresses are applicable for viewing information through web browsers. A common web browser is the Internet explorer through which internet is accessible.

7.1.2. HTTP and URL

The act of using a web browser is possible through HTTPs called HyperText Transport Protocol which hyperlinks with the World Wide Web address. For example, if we want to view the homepage of NCBI (National Center for Biotechnology Information) for which the website address is www.ncbi.nlm.nih.gov, it is accessed through http://www.ncbi.nlm.nih.gov. This enables the homepage to be viewed and browsed. This unique address is the Uniform Resource Locator or the URL which specifies the directory path and the filename (s) of the web address.

7.1.3. Softwares

Bioinformatics is the development of application softwares in the Life Sciences. Many numbers of softwares are being developed for the end-user in all profiles such as genomic data, protein structural information and drug discovery and development. These softwares are based on those being developed by the IT industry. Such softwares include JAVA, JavaScript, PERL and XML that have been extended for use in bioinformatics.

JAVA (from *java beans*) is a portable programming language with its source code based on C++ that can run on all hardware platforms, from large servers to individual PCs, without modification.

JavaScript is a scripting language using a similar syntax to Java, and is used primarily to enhance the World Wide Web pages with pop-up windows and animations.

PERL (Practical Extraction and Reporting Language) is a versatile scripting language widely used in bioinformatics for analysis of sequence data.

XML (eXtensible Markup Language) allows files to be described in terms of the types of data displayed on a WWW page and how the data is processed by other programs or by a database management system.

7.2. DNA and Protein Information Resources

DNA databases are classified into Primary databases and Specialized databases catering to the ongoing needs of research activities for genomic annotations and sequence analysis. We discuss the Primary DNA sequence databases that envisage the enormous applications in bioinformatics.

7.2.1. NCBI

NCBI is the abbreviation for National Center for Biotechnology Information located in the campus of National Institutes of Health (NIH) Bethesda, MD, and accessed from: http://www.ncbi.nlm.nih.gov

NCBI was established in 1988 as a division of National Library of Medicine. This database aims to provide automated systems for storing and analyzing biological information, develop advanced methods of computer-based information processing, facilitate user access to database repositories and softwares and co-ordinate efforts to gather biotechnology information worldwide. GenBank is the DNA database of NCBI which stores and retrieves biological data for DNA analysis. This database is managed by curators whose roles are to create sequence data records from scientific literature, submitted directly by authors and shared *pro rata* from EMBL and DDBJ.

7.2.2. EMBL

European Molecular Biology Laboratory has its headquarters in Heidelberg, Germany. DNA database management is provided by European Bioinformatics Institute, EBI, in the United Kingdom which is an outstation of the EMBL and can be accessed from: http://www.ebi.ac.uk

EBI is one of Europe's primary nucleotide sequence data resource and is a collaborative project with NCBI and DDBJ and ensures updated database entries on a daily *pro rata*. EBI also collaborates with Swiss Institute of Bioinformatics for the management of protein sequence information.

7.2.3. DDBJ

DDBJ is the DNA Data Bank of Japan and is in collaboration with EMBL and NCBI. This database produces, maintains and distributes DNA sequence information at the National Institute of Genetics and sequence submission is made possible from allover the world by a web-based data submission tool. DDBJ is accessed from: http://www.ddbj.nig.ac.jp

Protein information resources are classified into the Primary sequence databases, Secondary databases and structure databases for each architectural level of proteins which has enabled innumerous investigations on proteins which were earlier difficult to obtain only from experimental approaches. We present some information on the Primary protein sequence databases and few programs used for protein visualization and identification.

7.2.4. PIR

The first Protein sequence database was developed at the National Biomedical Research Foundation (NBRF) by Margaret Dayhoff for investigations of evolutionary relationships among proteins. Since 1988, it is maintained collaboratively by PIR at NBRF and JIPID, Japan and MIPS, Germany.

7.2.5. MIPS

The Martinsried Institute for Protein Sequences of the Max-Planck Institute is the European partner of the PIR to collect, distribute and maintain up-to-date protein sequences across Europe. MIPS also provide access to a database of aligned protein families and database search softwares for retrieval of homologs.

7.2.6. SWISS-PROT

SWISS-PROT was initiated from the University of Geneva and the EMBL in 1986 and moved to EBI in 1994. In 1998, it again moved to Swiss Institute of Bioinformatics (SIB) and since then it is maintained collaboratively by SIB and EBI/EMBL. The database endeavours to provide high-level annotations, descriptions of functions of proteins, domain structures and post-translational modifications.

Mala and Takeuchi (2008) have analysed the bioinformatics approaches for protein analysis (Tables 7.1, 7.2).

Table 7.1. Protein Visualization Programs

Program	Function
RasMol	3-dimensional visualization
Cn3D	3-dimensional visualization, linked to sequence alignments
Chime	3-dimensional visualization
TOPS	Visualization of protein folding topologies
DSSP	Finds secondary structure elements in an input structure
Surfnet	Visualization of protein surface
PROCHECK	Checks stereochemical quality of protein structures
PROMOTIF	Analyses protein structural motifs

Table 7.2. Protein identification and Characterization Programs

Program	Function
AACompIdent	Identification of amino acid composition
TagIdent	Identification of proteins using mass spectrometric data
PeptIdent	Identification of proteins using peptide mass fingerprinting data
MultiIdent	Identification of proteins using pI, MW, amino acid composition
Propsearch	Find putative protein family
PepSea	Identification of protein by peptide mapping or peptide sequencing
FindPept	Identification of peptides resulting from unspecific cleavage of proteins
TMAP; TMHMM	Prediction of transmenbrane helices
ProtParam	Computation of physical and chemical parameters of a protein

7.3. Database Structure and Analysis

7.3.1. BLAST

Basic Local Alignment Search Tool (BLAST) is a sequence analysis and retrieval tool for DNA and Protein query sequences from a DNA or protein database. BLAST finds regions of local similarities between biological sequences rather than a global fit. The search tool enables a rapid sequence input and retrieves sequence information to the user within a few minutes. This makes it a widely used sequence analysis search tool for both DNA and protein sequences. The basic BLAST consists of:

BLASTN searches a nucleotide database for a nucleotide query
BLASTP searches a protein database for a protein query
BLASTX searches a protein database for a translated nucleotide query
TBLASTN searches a translated nucleotide database for a protein query
TBLASTX searches a translated nucleotide database for a translated nucleotide query
PSI-BLAST and PHI-BLAST (Position Specific Iterated BLAST or Pattern Hit Initiated BLAST) algorithms for a protein query also function in the BLASTP.

7.3.2. Clustal W

Clustal W is a multiple sequence alignment program for DNA or protein sequences and is accessed from EBI database. A recent version is the Clustal W2 and follows a similar input of sequences as BLAST provided a minimum of two sequences have to be aligned. The default parameters are sufficient for a normal multiple alignment. Choice of parameters is also available. A multiple sequence alignment provides information on the conserved sequences among its family members or other species. Evolutionary relationships are also viewed as Phylograms or Cladograms.

7.3.3. DOMO

A new database of aligned protein domains known as DOMO has been developed by Gracy and Argos (1998). DOMO can be accessed through the sequence retrieval system (SRS). A form- based query manager allows retrieval of familial domain alignments by identifiers, sequence accession numbers or keywords. The DOMO sequence analysis provides a simple tool for determining domain arrangements, evolutionary relationships and key amino acid residues in a query protein sequence.

7.3.4. PROF-PAT

Bachinsky et al (2000) have developed Prof_Pat 1.3, an updated database of patterns to detect local similarities, containing patterns of more than 13,000 groups of related proteins in a format similar to that of PROSITE.

Figure 7.1 indicates a hierarchy of the databases used in protein analysis.

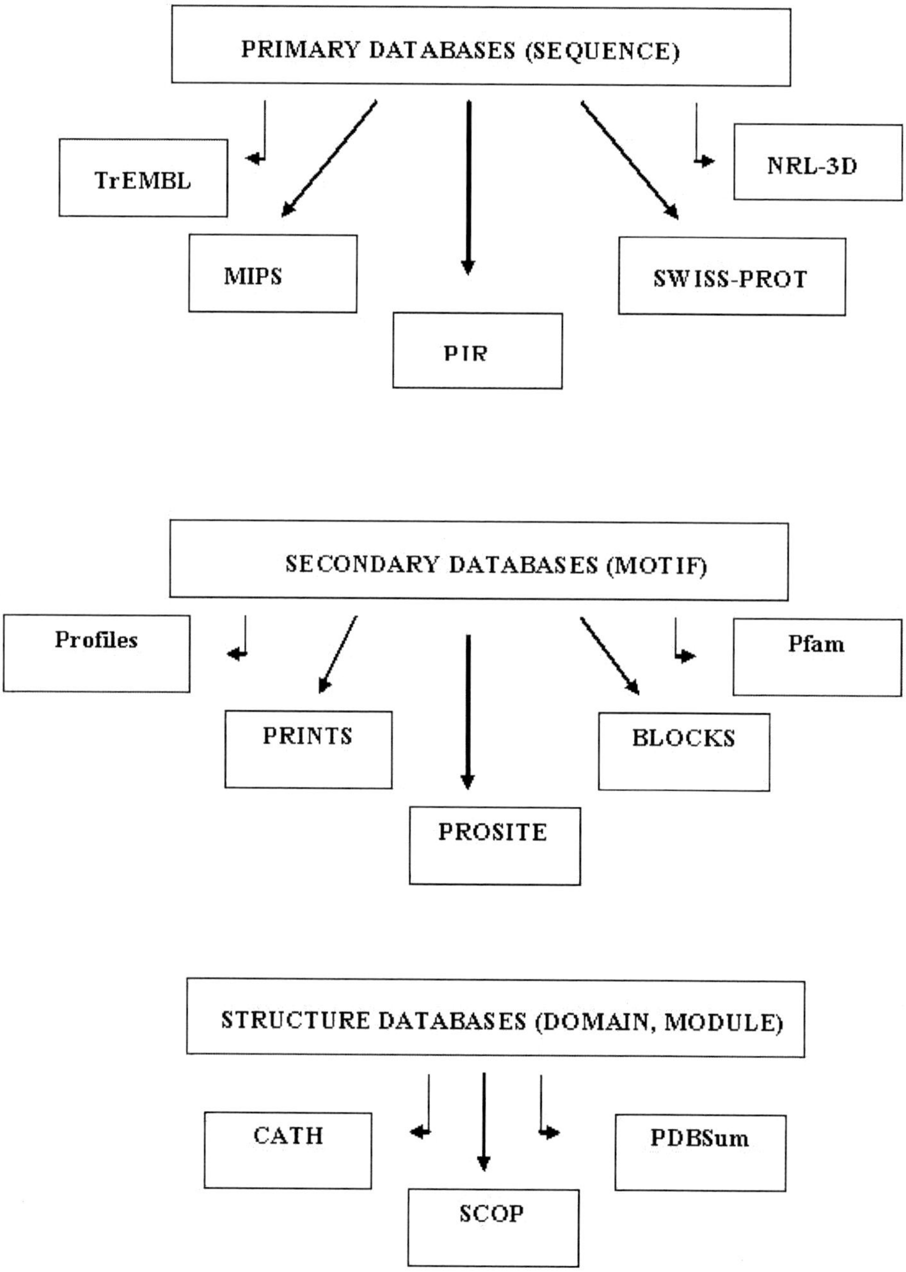

Figure 7.1. Database structure for protein sequence and structure analysis.

7.4. Lipase Engineering Database (LED)

Pleiss et al. (2000) have constructed a database exclusively on lipases to understand and exploit sequence-structure-function relationships. The LED is available at http://www.led.uni-stuttgart.de

Sequences of 92 microbial lipases and homologous serine hydrolases have been assigned to 32 homologous families and 15 superfamilies. Multisequence alignments of all homologous families and superfamilies have also been provided. Functionally relevant amino acids have been annotated. Lipase Engineering Database (LED) serves as a useful tool for protein engineering to help understand the functional role of individual amino acids by reference to annotated aligned sequences and superimposed structures of microbial lipases.

Self-evaluation

(1) BLASTX search tool is for a
(a) protein query (b) nucleotide query (c) translated nucleotide query
(d) all the above

(2) Which of the following softwares is used in bioinformatics
(a) PERL (b) Java (c) XML (d) all the above

(3) Which of the following is not a primary database
(a) NCBI (b) DDBJ (c) PIR (d) SGD

(4) A multiple sequence alignment tool is available from
(a) EBI (b) PDB (c) EMBL (d) SWISS-PROT

(5) Lipase engineering database is available for
(a) lipase sequence (b) lipase structure (c) both the above
(d) none of the above

Answers

1c 2d 3d 4a 5c

Chapter 8

LIPASES: BIOCATALYSTS FOR THE FUTURE

8.1. INTRODUCTION

BIOCATALYSTS are enzymes that have the properties of a chemical catalyst employed in a chemical reaction, but are of biological origin. A biocatalyst serves to catalyse reactions by exhibiting marked specificities to result in targeted products of action. There are several advantages of biocatalysts over conventional catalysts and their importance in biocatalysis is due to the Energy of activation (E_a). For any given catalysis, the E_a is of primary significance that predicts the rate of the reaction. Biocatalysts reduce the reaction rates by a reduction in the $E_{a.}$ This is a significant criterion of the reactions catalysed by enzymes.

Enzymology is a vast science and we attempt to describe only the importance of lipases as biocatalysts, their structural features that attribute towards their novel properties, explanations for their significance of catalysis, starting with an introductory note on the significance and properties of biocatalysts.

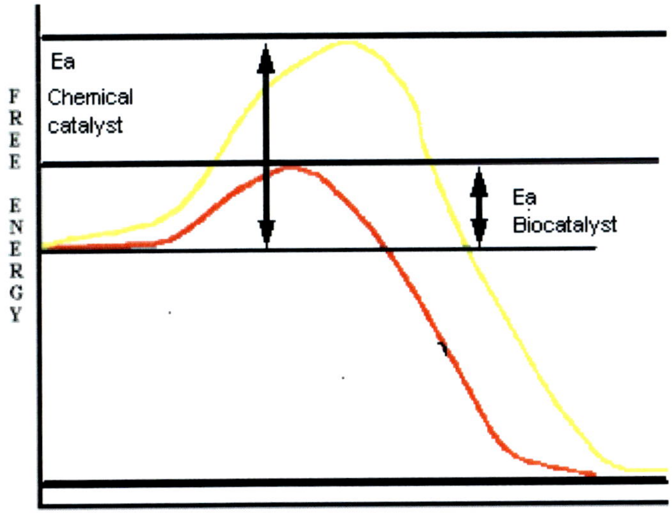

Figure 8.1. Activation energy of chemical and bio-catalysts.

The successful use of a biocatalyst depends largely on the market need of the biocatalyst for commercial purposes that have an increasing demand with the developments in lifestyle and industrial production processes. A marginal 75% of industrial enzymes are hydrolases and 90% of them are produced from microbial origin. The major source of biocatalysts is from nature and screening of such sources is deemed vital for any successful biocatalyst. Recent methods such as high-throughput screening could be coupled with sampling from extreme environmental niches. The biocatalyst thus obtained could be developed further for effective purposes by site-directed mutagenesis, fermentation optimization and gene cloning.

A critical factor is the efficacy and cost-effective procedures for integration of the biocatalysis step into the manufacture process. Screening of a suitable biocatalyst also relies on the useful properties of the biocatalyst to be employed in the industrial process. Screening is also of importance in identification of scientifically novel biocatalysts, but their useful properties could be better targeted. In essence, an ideal biocatalyst should possess certain criteria for marketing, such as, high volumetric activity (g/L/day); required selectivity; broad substrate specificity; rapid process definition; ease of process integration; cost-effectiveness; easy scale-up; validation and commissioning, reproducibility and reliability; patentability and marketing potentials. Another important consideration for the use of a biocatalyst is the costs of raw materials, capital equipment funding, market size of products and profit margins. With all these regards plus the use of cost-effective manufacturing process, proprietory technology and patentability, the investigations of the uses of biocatalysts and their improvements such as protein engineering to target desired products seem to prompt several researchers in the applied field of biocatalysis.

8.2. Lipase Biocatalysis

Lipases are a class of enzymes that have a significantly higher industrial potential in the enzyme market. The economic importance of lipases to the enzyme industry had earlier been restricted but has renewed interests since the past decades. The well-known specificity of enzyme-catalysed reactions coupled with regioselectivities of lipases has made the exploration of biocatalysts for industrial applications, with an attractive option compared with chemical methods, inspite of certain difficulties that could be altered, thereby increasing the demand in various kinds of applications. Recognition of the advantages of the use of lipases relative to traditional chemical processes indicates that lipases may be expected to gain importance in the enzyme market. Development of lipase technologies for the synthesis of novel compounds would probably target future impact in commercializations. The use of lipases is becoming increasingly important in a number of applications. Lipases have distinct advantages compared to classical enzyme catalysts: (a) ambient and mild reaction conditions; (b) no side reactivities; (c) ease of control of reactions; (d) unique specificities; (e) feasibility of desired reactions; (f) obtain new products with predetermined physical and chemical properties.

8.3. Lipase Catalysis in Organic Solvents

Non aqueous enzymology has emerged as a major area of biotechnology research and development. The understanding of the catalytic properties of lipases in non aqueous media is expanding rapidly and from a biotechnological perspective, there are innumerous advantages in using organic solvents which favor proper enzyme conformation. Reaction catalyses that are not possible in aqueous media can be performed with high end products in organic solvents. In organic media, lipases acquire remarkable properties with enhanced stability, tailor-made substrate and enantiomeric specificities and molecular memory. However completely anhydrous organic media cannot be favoured due to the water solubility of the enzyme that is essential for the catalysis. The rigidity of the enzyme structure at low water content imparts novel substrate specificities, pH memory and molecular imprinting. A thin water layer is therefore essential for an ideal catalysis in organic solvents. Thus the catalysis in organic media may be biphasic, microaqueous, or micellar. The nature of organic solvents is also an important criterion in deciding the reaction feasibility. Hydrocarbon solvents like hexane are immiscible in water and in the presence of an aqueous enzyme phase result in inherent stability of the lipase. However, with water-miscible solvents such as acetone, the essential water may be extracted from the proteinaceous backbone of lipase thereby depriving the catalytic action of the lipase. Enzymes such as lipases have been investigated for their promising potentials in organic solvents suited for various industrial applications.

8.4. Applications of Lipase Catalysis

Lipases are endowed with remarkable properties of catalyses that find immense industrial applications. The multiplexity of lipases is due to their reactions of hydrolysis, esterification, transesterification and interesterification (see Chapter 1, Box 1.4). These catalytic properties enable innumerous reactions that can be manipulated and commercialized. Several Companies have come up with several lipases suitable for various end applications. One of the first enzyme products developed is Lipozyme^{TM} from *Mucor miehei* by Novo Nordisk, Denmark, for transesterifications, with high regioselectivity and good thermostability. On immobilization on an anion-exchange resin, the enzyme proved the industrial requirements without loss of productivity or selectivity. A remarkable property of Lipozyme^{TM} is its half-life of 1800 h at a process condition of 60°C. We hereby provide a few applications of lipases in commercialization.

8.4.1. Lipases in Oleochemical Processing

The oleochemical processing industry is the mostly large user of lipases due to hydrolysis of oils and fats, esterification of fatty acids, transesterification and interesterification to produce PUFAs. In this context we mention a few of these processes using lipases.

Lipase-catalysed Hydrolysis of Oils and Fats

Hydrolysis of oils or fats by conventional methods result in dark-coloured fatty acid products and are energy-intensive. Lipase-catalysed hydrolysis results in specificity of fatty acid production and under ambient conditions. Fatty acids find several uses in the manufacture of soaps, lubricants, paper chemicals, candles and textile auxiliaries.

Lipase-Catalysed Esterification

Ester synthesis is a viable production process for several end product applications. Triglyceride esters of commercial significance have been targeted. Sorbitol esters and glycoside esters have been studied with commercial viabilities.

Lipase-catalysed Transesterification

A transesterification process may be acidolysis or alcoholysis to result in commercialization of the end products. Transesterification of castor oil, glycoside esters, and production of PUFA-esters are a few examples.

Lipase-catalysed Interesterification

Interesterification is a process of producing a desired or target triglyceride from less beneficial triglycerides. For example, butterfat is a highly saturated triglyceride and interesterification could to be used to modify to its desired component, by incorporation of unsaturated fatty acid molecules. Similarly, coconut oil is highly unsaturated for edible purposes and interesterification to modify with unsaturated components can be considered.

8.4.2. Lipases in Detergent Formulations

Application of lipases as detergent additives to remove lipid-stained fabrics has long been an interesting area of investigation. An ideal detergent enzyme should be stable at higher temperatures, alkaline pH and in presence of proteases. Lipases fulfil these criteria and could therefore be used as additives in detergent formulations along with proteases. A detergent lipase has been marketed in 1988 named Lipolase produced from an *Aspergillus oryzae* strain. The suitability of a detergent lipase can also be studied in the presence of surfactants that provide an oil-water interface for the action of lipase.

8.4.3. Applications of Lipase in 'Green Chemicals'

Green chemicals are new products that are difficult to synthesize by traditional means. Lipases can be used for these purposes which are built up from fatty acids and simple glucosides, and exhibit surfactant properties. An example is the use of *Candida antartica* B lipase for synthesis of glucoside esters. Green chemicals are environmentally acceptable due to biodegradability and lower toxicity levels.

8.5. Mechanisms of Lipase Catalysis

A typical environment in which a lipase functions consists of a lipid droplet that is separated from the aqueous milieu by a two-dimensional surface phase. This surface phase is visualized as a monolayer of proteins and amphipathic lipids. This monolayer phase comprises a negligible fraction of the total mass in the system. The instantaneous rate and direction of the reaction depend on the concentrations of reactants within the surface phase. These surface concentrations are determined through partitioning reactions between the two phases and the surface phase (the interface). Partitioning of the enzyme from the aqueous phase to the surface phase is usually distinct from the interaction of the enzyme with its substrate within the interface.

The 'interfacial activation' of lipase is its property of the ability to hydrolyse lipidic substrates and not aqueous substrates but taking place at the interface between the lipid phase and the aqueous phase in which the enzyme is dissolved. The interfacial activation phenomenon was first recognized in 1936 by Holwerda et al. and then in 1945 by Schonheyder and Volqvartz. Earlier assumptions to distinguish between lipases and esterases were based on the aggregation of substrates in water. In 1960, Desnuelle et al. postulated that a conformational change of lipases might be associated with interfacial activation, wherein the hydrophobic patch reveals the active site to enable hydrolysis. The first 3-D structure of a human pancreatic lipase was elucidated in 1990 as described in Nature. This suggested that the interfacial phenomenon might be due to an amphiphilic surface loop covering the active site of the enzyme in solution, just like a 'lid'. When contact occurs with a lipid/water interface, this lid might undergo a conformational rearrangement as the result of which the active site becomes accessible.

Lipases are α/β hydrolases which have a common catalytic mechanism of ester hydrolysis, after binding of the ester substrate. A first tetrahedral intermediate is formed by nucleophilic attack of the catalytic serine, with the oxyanion, stabilized by two or three hydrogen ions, the 'oxyanion hole'. The ester bond is cleaved and the alcohol moiety leaves the enzyme. Lastly, the acyl enzyme is hydrolysed. The nucleophilic attack by the catalytic serine is mediated by the catalytic histidine and aspartic acid or glutamic acid.

An important step in the study of functionally active lipase is the ability to distinguish between an esterase and a 'true' lipase. The presence of the lid-like structure has triggered the formulations of mechanistic models since its conception. The spatial movement of the lid, associated with catalysis exposes hydrophobic patches, which in the closed lid conformation are solvent inaccessible. The hydrophobic patches are essentially involved in the docking of the enzyme to the lipid-water interface. In contrast to the unique feature of lipases, esterases do not possess the lid structure and hence are unable to catalyse lipid substrates high in long-chain fatty acids. However, esterases catalyse lower chain fatty acid triglycerides such as triacetin, tributyrin etc.

Candida rugosa lipase (CRL) is an extensively used biocatalyst and significant numbers of new processes and new applications for crude commercial (stabilised with lactose) mixtures have been developed, addressing production of NSAIDS, sterols modifications, improvements/applications with ionic liquids and other biocatalytical aspects. CRL is made up of Lip 1 to Lip7 isoenzymes, out of which Lip1, Lip2 and Lip3 are well-characterized and each isoenzyme shows different biocatalytic properties. Lip1 shows the strongest affinity for long-chain esters (a lipase behaviour) while Lip2, Lip3 prefer the hydrolysis of short-chain water-soluble esters (an esterase behaviour). Lip2 shows a higher activity than Lip3 towards cholesteryl oleate. This difference in properties could be related to a variation of the acyl chain in the cholesterol esters, while Lip3 shows better activity in the hydrolysis of p-nitrophenyl esters.

8.6. Interactions in Lipase Catalysis

Lipases have a specific geometry of binding sites for their substrate molecules. Based on binding site structures, lipases are classified into: (a) lipases with a hydrophobic, crevice-like binding site located near their protein surface (*Rhizopus* and *Rhizomucor* lipases), (b) lipases with a funnel-like binding site (*Pseudomonas* lipase) and (c) lipases with a tunnel-like binding site (*Candida rugosa*). The length of the scissile fatty acid binding site is between 7.8 A^o to 22 A^o.

Binding of substrates has been modeled for lipases from *Rhizopus*, *Rhizomucor miehei* and *Humicola lanuginosa*. The scissile fatty acid binding site has been studied using mutants with altered fatty acid chain length profile. The catalytic machinery is located at the bottom of a deep, elliptical binding pocket, where the ester substrates bind with its acid-alcohol axis parallel to the long axis of the binding pocket. Lipases have a long hydrophobic scissile fatty acid binding site located inside the binding pocket. Length and hydrophobicity of this binding site correlates with the fatty acid chain length profile of the respective lipases. Hence, the chain length specificity could be predicted from the structure of the lipase. This happens when the substrate is shorter than the scissile fatty acid binding site inside the binding pocket. As the chain length exceeds the binding pocket, the conformation of the fatty acid chain cannot be predicted. This occurs when the fatty acid may bind to the hydrophobic protein surface or stick into the hydrophobic substrate interface or even fold back into the binding pocket.

CRL shows enhanced enantioselectivity than other microbial sources due to the conformation of its active site. The active site is located much deeper in the protein interior. The presence of a tunnel-shaped acyl binding site in *Candida rugosa* and its larger size creates a scaffold where substrate/protein interactions may form along the entire acyl moiety of the substrate. This explains the higher enantioselectivity towards chiral acyl moieties than lipases with more superficial active sites.

Lipases show differences in steric and electronic environments around their active site and as a consequence, different substrate selectivities. The structural features of the active site determine the extent of lipid interaction with the hydrophobic substrate-binding domain, the accessibitlity of the enzyme for the substrate and the solvent, and the selectivity of substrate-protein interaction. Based on investigations of lipase-inhibitor complexes, it has been shown that the accessibility of the lipid-binding site was lowest in water. This indicates that solvent polarity has a function in the conformational status of a lipase and the feasibility of its reactions.

One of the factors explaining lipase activity is the contribution of electrostatic interactions. The electrostatic potential distribution on the molecular surface of the lipase is a function of pH and the presence and distribution of polar and hydrophobic residues in the active cleft. The 'electrostatic catalpult model' has been proposed by M.T.N. Peterson, P. Fojan and S.B. Peterson (2001) for the negative potential in the active site which is correlated with maximum activity towards triglycerides. The study of electrostatic potential distribution could also be applied for metal binding sites and/or allosteric region of an enzyme, although in the case of microbial lipases, metals are not required as cofactors.

8.7. Lipase Protein Design for Enantioselectivity

Enantioselectivity is the property of lipases that exhibit chiral specificity. Methods to increase the selectivity for distinguishing between enantiomers is a novel approach towards protein design or protein engineering which can be achieved by functional expression of isoenzymes, rational computer-guided protein design and directed evolution.

8.7.1. Functional Expression of Isoenzymes

Commercial grade lipases are sold as crude products, their activity and selectivity reflecting only a weighted average of the properties of various isoenzymes and other enzymatically active impurities. Enantioselectivity in these cases may change from batch to batch depending on the operational stability of isoenzymes. In the case of CRL, cross-linked crystals showed much higher enantioselectivity than their commercial preparations in the resolution of ketoprofen. This increase in enantioselectivity was attributed to the selective enrichment of one isoenzyme during crystallization. The commercial preparations show only minor selectivity (E=15) while recombinant CRL shows excellent enantiomeric purity (E > 100).

8.7.2. Rational Computer-guided Protein Design

The use of computer modeling to rationalize the substrate specificity and enantioselectivity has earlier been approached. An improved enantioselectivity from E=1.6 in wild type to E=22 in Thr40 → Val mutants has been observed. The stereoselectivity of lipase from *Pseudomonas cepacia* towards secondary alcohols and γ- and δ- lactones has been modeled.

8.7.3. Directed Evolution

Directed evolution represents an efficient alternative to rational protein design when the structure and mechanism of the targeted enzymes are not available. This method consists of the generation of large mutant libraries by random mutagenesis and recombination followed by advanced high-throughput screening. Mutants have also been identified having opposite stereopreferences. Recently, a combinatorial approach using cassette mutagenesis has resulted in E values above 50.

These protein engineering approaches create an understanding of the basis of enhancement of enantioselectivity. Another approach is the substrate and medium engineering by variations of the solvent system to influence enantioselectivity. This aspect is beyond the scope of this book and is therefore not detailed.

The use of biocatalysts has evinced keen interests among chemists and biochemists, and with lipases which are promising candidates, there are innumerable applications in the current scenario. For example, use of lipases can be of interest in immobilized forms suitable for application in forming bioreactors and in preparations as biosensors for clinical applications. With protein engineering of lipase, it can be brought into considerations for the degradation of cholesterols *in vivo* by targeted enzyme specificities. Protein conformations are important in predicting the feasibility of biocatalysis, and hence a deeper understanding of the conformational states of lipase catalysis could yield newer applications.

Self-evaluation

(1) Biocatalysts
(a) lower the Ea (b) increase the Ea (c) have no effect on Ea (d) none of the above

(2) Lipase catalysis is
(a) homogeneous (b) heterogeneous (c) none (d) both

(3) Synthesis of methyl oleate is
(a) esterification (b) transesterification (c) Interesterification (d) all the above

(4) A characteristic of non aqueous enzymology is
(a) pH memory (c) enantiospecificity (c) enhanced enzyme stability (d) all the above

(5) A lipase structural feature is
(a) α/β hydrolase fold (b) lid (c) tryptophan hydrophobic residue (d) all the above

Answers

1a 2b 3a 4d 5d

Section I

EXPERIMENTS

I. SPECTROPHOTOMETRIC LIPASE ASSAY

There are many approaches for the determination of lipase activities, while the selection of a suitable assay procedure is a prime criterion in use or study of their applications/characteristics. For example, isolation methods require rapid assays without sacrifice of sensitivity and generally spectrophotometric assays are preferred. A usual spectrophotometric assay is the pNPP (p-nitriphenypalimitate) method where the colored p-nitrophenol in alkaline condition is measured. However, this method presents a wrong assumption of the esterase activity also which is significantly error-based. Hence this method can be used during purification protocols of lipase, where true lipases are employed. Radiolabelled substrates can be used for lipase activity determinations for study of their characteristics and structure-functions. HPLC and software-based assays offer high sensitivity and accuracy but require cost input and is not common for routine lab practices. Titrimetry is manually cumbersome but again lacks sensitivity due to non-clarity of end points and is very much prone to error especially during isolation methods of a true lipase producer that does not identify little amounts of lipase activities. A good and reliable lipase assay would be to use a spectrophotometric protocol with speed, accuracy, sensitivity and this has been devised by Kwon and Rhee in 1986 using a copper reagent for colored complexation of the FFAs and extraction in an organic solvent. The solvent phase is measured in the visible region and correlates to the lipase activity of the enzyme preparation. A working modification of the protocol is presented here.

Reagents

Substrate

An emulsion of 37.5 ml olive oil and 12.5 ml 2% PVA (polyvinyl alcohol, MW 14,000, heated to dissolve in distilled water by rapid stirring) is mixed by homogenization for 10–15 min. This is the substrate emulsion for the assay procedure.

Buffer

0.1 M Tris.HCl, pH 7.0

6 N HCl

Isooctane

Copper Reagent

5% cupric acetate (w/v) is dissolved in distilled water and pyridine is added to pH 6.1 and madė up to 100 ml. This reagent is stable and can be stored for a month at RT.

Protocol

1.0 ml of substrate emulsion and 0.8 ml of buffer are preincubated at 37°C for 10 min followed by addition of 0.2 ml of enzyme and incubated at 37°C for 20 min. 1 ml of 6 N HCl is added to stop the reaction and 1.5 ml of isooctane is added to extract the FFAs liberated by rapid vortex of the mixture. The isooctane phase is separated from the reaction mixture, added 0.5 ml of copper reagent and vortexed to extract the Cu-FFA. The solvent phase containing the blue complex is measured at 715 nm. A blank is run simultaneously without the enzyme and compensated by buffer.

Standard

100 μM-1 mM of a 10 mM oleic acid stock dissolved in isooctane is used as the standard. Addition of 0.5 ml of copper reagent and subsequent absorbance of the isooctane phase at 715 nm is followed. A standard graph is plotted with the absorbance readings on the Y-axis and the concentration of oleic acid on the X-axis. The unknown concentration of lipase is derived from the calibrated plot.

Calculation

Lipase activity is determined from the concentration obtained and taking the dilution/amount factor and time of incubation of the enzyme added to the reaction mixture.

Lipase Unit Definition

Lipase activity is defined as the amount of enzyme that liberates 1 μmol of FFAs under standard assay conditions. Lipase activity is expressed as Units/ml/min.

II. Protein Estimation by Bradford's Method

Protein estimation is a significant unit of expression of enzyme/protein under study. This value is an estimate of the amount of the particular protein content among the whole of the protein mixture and is defined in terms of specific activity. Specific activity is the amount of the desired protein present per volume of the protein and is expressed as Units per mg.

Specific activity = Enzyme activity (U/ml) / protein content (mg/ml) U/mg

There exist several methods to determine the protein content with respect to sensitivity. The most widely used protein estimation is the Lowry's method suitable for clinical investigations. Biuret method is a crude method of estimation, common in educational lab practice. Bradford's method is a good choice for any protein of interest. This method is widely acceptable and would be suitable for purification procedures that require simple and rapid methods. The reagent is stable and can be stored for about 6 months in cold conditions. Disposable glass cuvettes may be used since the blue complex adheres to the walls of the cuvettes. Quartz cuvettes may be used but have to be rinsed with 6 N HCl or 40% methanol.

Reagents

Bradford's Reagent (5X)

100 mg of Coomassie Brilliant Blue G-250 is dissolved in 50 ml of 95 % ethanol. 100 ml of conc. Phosphoric acid is added and a final volume of 200 ml by addition of distilled water is prepared.

Protocol

Protein samples are made up to 1 ml with distilled water and 1 ml of 1X Bradford's reagent is added and vortexed well. The protein content is measured at 595 nm before 1 h. A reagent blank is used for the measurements.

Standard

20-200 mg of BSA solution (200 mg BSA is prepared in 100 ml standard flask with addition of NaOH pellets sufficient to dissolve the BSA) is used as the standard and a calibrated plot is obtained. The unknown protein content is calculated from the standard graph with conversion to per ml of undiluted enzyme (calculated from dilution factor).

III. Recovery of Proteins in Presence of Lipids and Detergents

Dilute protein samples with salts, detergents, phospholipids, β-mercaptoethanol is difficult for protein estimations and hence recovery of concentrated protein samples without the interfering substances is required in day-day laboratory practice during routine protein purification procedures. The method of Wessel and Flugge (1984) is applicable for soluble as well as hydrophobic proteins of interest.

An aliquot of 0.4 ml of methanol is added to 0.1 ml of protein sample and the samples are vortexed to mix well and centrifuged at 8000 rpm for 10s for total collection of the sample. 0.1 ml of chloroform is then added, vortexed and centrifuged at 8000 rpm for 10s. Protein samples containing phospholipids require addition of 0.2 ml of chloroform. 0.3 ml of water is added for phase separation with vigorous vortexing and centrifuged at 8000 rpm for 1 min. The upper phase is discarded. 0.3 ml of methanol is added to the lower chloroform phase and the interphase containing the precipitated protein. The samples are mixed and centrifuged at 8000 rpm for 2 min to pellet the protein. The supernatant is removed and the protein pellet is air-dried and stored until use.

For protein estimation, the pellets are solubilized by addition of 50–100 μl of 5% SDS for Lowry method. For SDS-PAGE, the protein pellets are dissolved in 50 μl of sample buffer containing 5% SDS.

This method can be scaled-up to larger volumes of protein samples in which case, 1.0 ml of proteins can be precipitated with 4.0 ml methanol, 1.0 ml chloroform and 3.0 ml water. After centrifugation and removal of the upper phase, the lower phase and interphase are transferred to 1.5 ml tubes with subsequent procedures as described.

This procedure avoids heating or acid conditions for the protein precipitation and hence is useful for removal of protein from extracts which are labile to extreme pH or heat. The method is also effective for removal of lipids from membranes with high lipid/protein ratio.

IV. Protein Detection in Polyacrylamide Gels without Staining

Protein visualization is generally done by Coomassie Brilliant Blue (CBB) and silver staining protocols in common lab practice. Use of conventional CBB method requires repeated handling of the toxic gels, longer periods for staining and destaining and consumption of large volumes of methanol and acetic acid. A modified CBB protocol is available that uses rapid staining but however produces objectionable odours. Silver staining is another procedure and requires continuous monitoring of the gel with frequent gel handling. A report in Analytical Biochemistry (2004) describes the use of 2,2,2-Trichloroethanol (TCE) for fluorescent detection of proteins in a PAGE gel. The principle of this method is based on the presence of Trp in proteins that fluoresce in the visible range with trihalocompounds by an uv-induced reaction in a transilluminator and enable visualization.

0.5% (v/v) TCE is added to the gel preparation by dissolving in gel buffer and then acrylamide, SDS, APS and TEMED are added as per usual procedures. In another protocol,

12% SDS-PAGE gel is soaked in 10% TCE in water:methanol (1:1) for 10 min. The gel is then washed in water. Visualization of the protein bands is carried out by placing the gel on a 300 nm-UV transilluminator and irradiated for 2–5 min. Bluish-green bands against a pale blue gel matrix is observed.

This method is also compatible with 2-D PAGE, native PAGE gels, Western blotting and autoradiography.

V. Staining of Glycoproteins

Glycoproteins can be stained by Periodic acid-Schiff reagent which is useful for identification of glycosylated proteins.

After SDS-PAGE, the protein gel is washed continuously with 500 ml of 40% methanol and 7% acetic acid O/N. The gel is then soaked in 7.5% acetic acid and kept at RT for 1 h. The gel is then transferred to a gel tank containing 1% periodic acid and kept in dark for 1 h at 4°C. The gel is then washed in 7.5% acetic acid for 10 min with repeated washing for 6 times. Finally the gel is incubated in Schiff's reagent for 1 h in dark at 4°C, washed in 0.5% sodium metabisulfate and preserved in 7.5% acetic acid.

VI. Isolation of Total RNA

Isolation of total RNA is the primary step for RT-PCR where the mRNA species acts as the template for complementary DNA synthesis, based on which the DNA may be amplified using a set of primers. Isolation of RNA is critical and requires careful handling especially contamination from RNases which may cleave the template. Use of hand gloves and DEPC-treated materials is mandatory. Glasswares and other materials should be previously sterilized. The isolated RNA is then assessed for its purity: A_{260}/A_{280} should be 1.8-2.0. The quantification of RNA is based on the formula: 1 O.D at A_{260} = 40 µg RNA.

Bacterial cells are ground in liquid nitrogen in an autoclaved prechilled mortar and pestle. The cell mass is then mixed with 5 ml of RNA extraction buffer: 100 mM Tris.HCl, pH 8.0, 10 mM EDTA, pH 8.0, 1% SDS and 100 mM lithium chloride. 5 ml of water-saturated acidic phenol (previously prepared by repeated saturation) is added at 80°C and vortexed for 30s. An equal volume of 24:1 v/v of chloroform: isoamyl alcohol is added and vortexed again for 30s. The resulting emulsion is centrifuged at 10,000 rpm for 15min. The upper aqueous phase is removed and the RNA is precipitated with an equal volume of 4 M lithium chloride at -70°C for 1h. The RNA pellet is collected by centrifugation at 10,000 rpm for 15 min at 4°C and washed with 70% ethanol and dried under vacuum. The dried RNA pellet is dissolved in DEPC-treated water and stored at -70°C prior to use. An electrophoretic run is performed to detect the integrity of the RNA preparation. RNAse-free DNase I (0.1 U/µg) is added in case of genomic DNA contamination. The reaction is carried out at 37°C for 1 h in the presence of 0.4 U RNasin and heated at 65°C for 15 min. RNA is precipitated using 0.1 volume of 3 M sodium acetate buffer, pH 5.2 and 3 volumes of absolute ethanol at -70°C for 1 h. The RNA pellet is collected by centrifugation at 12,000 rpm for 10 min at 4°C and dried under vacuum. Total RNA is resuspended in 5 µl of DEPC-treated water.

VII. Formaldehyde Gel Electrophoresis of Total RNA

Formaldehyde electrophoresis is performed for total RNA/mRNA to check the integrity of RNA preparations. The gel apparatus has to be thoroughly cleaned prior to the electrophoresis run by soaking in 2-3 M NaOH O/N. The following day, it is thoroughly rinsed with 3–4 times distilled water and air-dried.

Reagents

The reagents for RNA electrophoresis are prepared previously:

5X Gel Running Buffer (MOPS Buffer)

0.1 M MOPS (3-(N-morpholino)propanesulfonic acid), pH 7.0
40 mM sodium acetate
5 mM EDTA, pH 8.0

20.6 g of MOPS is dissolved in 800 ml of DEPC-treated 50 mM sodium acetate. Adjust the pH to 7.0 with NaOH. Add 10 ml of DEPC-treated 0.5 M EDTA. Adjust the volume of the solution by adding water and filter-sterilize with 0.2 μ Millipore filter.

Loading Buffer (100 μl)

48 μl of deionized formamide
17.3 μl of 37% formaldehyde
34.7 μl of loading dye

Loading Dye

160 μl of 10X MOPS-treated water
100 μl of DEPC-treated water
100 μl of ethidium bromide (10 mg/ml)
80 μl sterile glycerol
80 μl saturated bromophenol blue in sterile water

DEPC-treated Water

0.1% DEPC is added to water, left for 12 h and autoclaved.

Protocol

50 ml of agarose gel is prepared by melting 0.5 g of agarose in 42.5 ml of milli Q water. 10 ml of 5X MOPS is added and cooled to 55°C. 2.7 ml of 37% formaldehyde is added, mixed and poured into the gel cassette with placement of the gel comb and allowed to solidify by keeping at RT. The cooled gel is covered with 1X MOPS buffer (running buffer) and prerun at 50 V. 15-20 μg of RNA is mixed in 5 μl of loading buffer and boiled for 2 min and snap-chilled and is immediately loaded. The gel is run at 100 V and is monitored when the samples have run halfway through the gel in a gel documentation system. The integrity of

RNA bands are checked for 5 S, 16 S and 23 S for prokaryotic RNA and for 5S, 5.8 S, 18 S and 28 S for eukaryotic RNA. In case the gel bands are not separated, the gel is placed in the gel cassette and electrophoresis is resumed.

VIII. Vector Dephosphorylation

Vector dephosphorylation is an essential step towards a ligation protocol. Dephosphorylation prevents religation of cohesive ends and is desirable for a restriction digest that generates non-compatible ends.

Calf Intestinal Alkaline Phosphatase (CIAP)

Dilute sufficient CIAP for immediate use in 1X reaction buffer to a final concentration of 0.01 u/μl. Each pmol of DNA ends will require 0.01 u CIAP (added in two aliquots of 0.005 u/pmol ends).

Vector DNA

1 μg of 1kb DNA fragment equals approx.1.5 pmol DNA or 3 pmol ends. The DNA should be purified by ethanol precipitation. The DNA pellet is resuspended in 40 μl of 10 mM Tris-HCl, pH 8.0 or water.

Reaction Mixture

DNA (upto 10 pmol of 5′-ends)	40 μl
CIAP 10X reaction buffer	5 μl
Diluted CIAP (0.01 u/μl, add 0.005 u/pmol ends)	upto 5 μl
Deionized water to a final volume of	50 μl

CIAP Stop Buffer

10 mM Tris.HCl, pH 7.5
1 mM EDTA, 200 mM NaCl
0.5% SDS

Ethanol Precipitation

Add 0.5 vol of 7.5 M ammonium acetate, pH 5.5 and 2 vols of 100% ethanol to the final aqueous phase.

Reaction Protocol

For 5′-protruding ends, incubate at 37°C for 30 min. For 5′-recessed or blunt ends, incubate at 37°C for 15 min and then at 56°C for 15 min. Add another aliquot of diluted CIAP equivalent to the concentration in reaction mixture and repeat the incubation. Add 300 μl CIAP stop buffer and follow phenol/chloroform extraction and ethanol precipitation.

IX. Transformation by PEG Method

Chung, Niemela and Miller (1986) have developed a simple and rapid method of transformation by a one-step PEG method for bacterial cells.

LB Broth (w/v)

Tryptone	1.0%
Yeast extract	0.5%
Sodium chloride	1.0%
Water	100 ml

2X TSS (Ice-cold)

LB broth
10% PEG (MW 3350-8000)
DMSO 5% w/v
20-50 mM $MgSO_4$ or $MgCl_2$
pH 6.5

Dilute sufficiently (1:100) a fresh O/N bacterial culture with a pre-warmed LB broth and incubate in a rotary shaker to an O.D of 0.3–0.4. Add an equal volume of ice-cold 2X TSS and mix gently. For transformation, 0.1 ml aliquot is pipetted into a cold polypropylene tube containing 1 µl (100 pg) of plasmid DNA and cell-DNA suspension is mixed gently. This is incubated for 5–60 min at 4°C. A 0.9 ml aliquot of TSS (or LB broth) and 20 mM glucose is added and the cells are incubated at 37°C at 225 rpm for 1 h to allow expression of Ab^r gene and the transformants are selected by standard methods. For transformation of lipase gene, it is ideal to avoid glucose since lipases are subjected to catabolite repression and identification of transformants using tributyrin plate assays may be ineffective. For long time storage of the TSS competent cells prior to transformation, the cells are frozen in dry ice/ethanol bath and stored at -70°C until transformation and thawed slowly on ice and used immediately. Transformation efficiency is the number of transformants per µg input plasmid DNA/10^8 cells.

X. Preparation of Surfactant-coated Lipase

Surfactant-coated lipases enable lipase catalysis in organic solvents that have tremendous applications in organic syntheses and oleochemical processing. Surfactant-coated lipases acquire altered specificities, pH memory and are useful candidates in non-aqueous enzymology.

Preparation of Surfactant-Modified Lipase by Basheer et al. (1995)

Crude lipase is dissolved in 1L Tris buffer, 0.1M, adjusted to pH 5.0 and stirred magnetically at 4°C. Sorbitan monostearate (0.75g) dissolved in 20 ml of ethanol is added dropwise into the stirred enzyme solution. The mixture is sonicated for 15 min and stirred for 2h at 5°C. The precipitate is collected by centrifugation, frozen at -20°C and freeze-dried. The yellowish solid is collected after freeze-drying.

Preparation of Surfactant-coated Lipase by Kamiya et al. (1995)

One gram per L lipase in 500 ml of 0.1M phosphate buffer solution, pH 6.9 and 0.5g of surfactant is mixed and sonicated in an ultrasonic bath for 20 min. The transluscent solution is incubated for 24 h at 4°C and collected by centrifugation and dried under reduced pressure (vacuum drying). A white powder is obtained with 20% yield. The enzyme content of the surfactant-coated lipase is calculated from the results of elemental analysis of the lipase and the lipase-surfactant complex.

APPENDIX

I. PREPARATION OF BUFFERS (BUFFER CHART)

A. 0.1 M Acetate Buffer

Prepare 0.1 M solution of anhydrous sodium acetate. Add acetic acid using a pipetteman or pipette, mix well using a glass rod and adjust to required pH using a pH meter or a pH probe. The pH range of acetate buffer is 3.6–5.6.

B. 0.1 M Phosphate Buffer

Prepare 0.2 M of NaH_2PO_4 (A) and Na_2HPO_4 (B). Add x ml of A and y ml of B, and dilute to 200 ml.

A (x ml)	B (y ml)	pH
93.5	6.5	5.7
92.0	8.0	5.8
90.0	10.0	5.9
87.7	12.3	6.0
85.0	15.0	6.1
81.5	18.5	6.2
77.5	22.5	6.3
73.5	20.5	6.4
68.5	31.5	6.5
62.5	37.5	6.6
56.5	43.5	6.7
51.0	49.0	6.8
45.0	55.0	6.9
39.0	61.0	7.0
33.0	67.0	7.1

28.0	72.0	7.2
23.0	77.0	7.3
19.0	81.0	7.4
16.0	84.0	7.5
13.0	87.0	7.6
10.5	90.5	7.7
8.5	91.5	7.8
7.0	93.0	7.9
5.3	94.7	8.0

C. 0.1 M Tris.HCl Buffer

Prepare 0.1 M Tris base solution. Add conc.HCl to the Tris solution using a pipetteman in small aliquots to the desired pH using a pH meter or pH probe. Carefully mix the Tris.HCl solution well before monitoring the pH as conc. HCl may corrode the pH bulb and affect the sensitivity. The pH of Tris base is ~ 11.0. Addition of higher volume of conc.HCl will drastically bring the pH to 2.0-3.0. Hence avoid using a pipette. The pH range for Tris.HCl can be used ideally to 6.0-9.0

II. Ammonium Sulfate Precipitation Table

For concentration of protein solutions by ammonium sulfate precipitation, small amounts of ammonium sulfate has to be added to the protein mixture, dissolved under cold conditions by gradual increasing concentrations and centrifuged to obtain a precipitate of the desired protein component of higher purity at each stage of addition. The following table represents the amounts of salt added to obtain 100% concentration from 0% concentration.

III. SDS-PAGE

A. Trichloroacetic Acid (TCA) Precipitation

TCA precipitation is required for performing an SDS-PAGE protocol for the removal of salts from the sample mix in any purification step after ammonium sulfate precipitation to avoid interferences by the salt which might produce diffuse bands on the gel.

	Final concentration of ammonium sulfate, % saturation at 0°C																
Initial conc. of ammonium sulfate (%)	20	25	30	35	40	45	50	55	60	65	70	75	80	85	90	95	100
	Gram solid ammonium sulfate to add to 100 ml of solution																
0	10.7	13.6	16.6	19.7	22.9	26.2	29.5	33.1	36.6	40.4	44.2	48.3	52.3	56.7	61.1	65.9	70.7
5	8.0	10.9	13.9	16.8	20.0	23.2	26.6	30.0	33.6	37.3	41.1	45.0	49.1	53.3	57.8	62.4	67.1
10	5.4	8.2	11.1	14.1	17.1	20.3	23.6	27.0	30.5	34.2	37.9	41.8	45.8	50.0	54.5	58.9	63.6
15	2.6	5.5	8.3	11.3	14.3	17.4	20.7	24.0	27.5	31.0	34.8	38.6	42.6	46.6	51.0	55.5	60.0
20	0	2.7	5.6	8.4	11.5	14.5	17.7	21.0	24.4	28.0	31.6	35.4	39.2	43.3	47.6	51.9	56.5
25		0	2.7	5.7	8.5	11.7	14.8	18.2	21.4	24.8	28.4	32.1	36.0	40.1	44.2	48.5	52.9
30			0	2.8	5.7	8.7	11.9	15.0	18.4	21.7	25.3	28.9	32.8	36.7	40.8	45.1	49.5
35				0	2.8	5.8	8.8	12.0	15.3	18.7	22.1	25.8	29.5	33.4	37.4	41.6	45.9
40					0	2.9	5.9	9.0	12.2	15.5	19.0	22.5	26.2	30.0	34.0	38.1	42.4
45						0	2.9	6.0	9.1	12.5	15.8	19.3	22.9	26.7	30.6	34.7	38.8
50							0	3.0	6.1	9.3	12.7	16.1	19.7	23.3	27.2	31.2	35.3
55								0	3.0	6.2	9.4	12.9	16.3	20.0	23.8	27.7	31.7
60									0	3.1	6.3	9.6	13.1	16.6	20.4	24.2	28.3
65										0	3.1	6.4	9.8	13.4	17.0	20.8	24.7
70											0	3.2	6.6	10.0	13.6	17.3	21.2
75												0	3.2	6.7	10.2	13.9	17.6
80													0	3.3	6.8	10.4	14.1
85														0	3.4	6.9	10.6
90															0	3.4	7.1
95																0	3.5
100																	0

Protocol

Add 0.5 ml of ice-cold 1.0% TCA (100 μl in 10 ml) to 0.1 ml sample prior to running the SDS-PAGE. Mix well by vortexing. Leave it in ice for 15 min. Microfuge and discard the supernatant. To remove TCA, add 0.5 ml of ice-cold 100% acetone. Vortex and leave it in ice for 15 min. Microfuge for 15 min. and discard the supernatant. The sample is now ready for processing by SDS-PAGE after addition of sample buffer containing SDS and β-ME.

B. Troubleshooting

PAGE gels may be misrun occasionally inspite of careful steps and proper reagents. Hence, to overcome these difficulties, there has to be some guidance to a successful run.

Gels Fail to Polymerize

Reagents should be at room temperature
Ammonium persulfate (APS) to be prepared freshly
Degas solutions longer
Check APS and TEMED additions
Ethanol inhibits polymerization. Do not wash glass plates with ethanol.

Electrophoresis Time is too Long

Check power supply current/voltage settings and wire connections
Check pH and concentration of the gel and tank buffers
Compare voltage with previous electrophoretic run
Check for tank buffer leaking

IV. Matrices for Chromatography

A. Gel Filtration Media

Matrix	**Fractionation range (Da of proteins)**
Sephadex# (Dextran)	
G-10	0–700
G-15	0–1500
G-25	1000–5000
G-50	1500–30,000
G-75	3000–80,000
G-100	4000–150,000
G-150	5000–300,000
G-200	5000–600,000

Bio-Gels* (Polyacrylamide)	
P-2	100–1800
P-4	800–4000
P-6	1000–6000
P-10	1500–20,000
P-30	2500–40,000
P-60	3000–60,000
P-100	5000–100,000
P-150	15,000–150,000
P-200	30,000–200,000
P-300	60,000–400,000
Sephacryl# (Dextran-polyacrylamide)	
S-100 HR	-
S-200 HR	5000–250,000
S-300 HR	10,000–1,500,000
S-400 HR	20,000–8000,000
Agarose	
Sepharose# 6B	10,000–4,000,000
Sepharose 4B	60,000–20,000,000
Sepharose 2B	70,000–40,000,000
Biogel* A-0.5	10,000–500,000
Biogel A-1.5	10,000–1,500,000
Biogel A-5	10,000–5,000,000
Biogel A-15	40,000–15,000,000
Biogel A-50	100,000–50,000,000
Biogel A-150	1,000,000–150,000,000

Pharmacia
* Bio-Rad

B. Ion-exchange Resins

Resin	Functional group	Matrix	Class
Anion exchangers			
AG 1	Tetramethyl ammonium	Polystyrene	Strong
AG 3	Tertiary amine	Polystyrene	Weak
DEAE-Sephacel	Diethyl aminoethyl	Sephacel	Weak
PEI-cellulose	Polyethyleneimine	Cellulose	Weak
DEAE-Sephadex	Diethylaminoethyl	Dextran	Weak
QAE-Sephadex	Diethyl-(2-hydroxyl-propyl)-aminoethyl	Dextran	Strong

Cation exchangers			
AG 50	Sulfonic acid	Polystyrene	Strong
Bio-Rex 70	Carboxylic acid	Acrylic	Weak
CM-Sephacel	Carboxymethyl	Sephacel	Weak
P-cellulose	Phosphate	Cellulose	Intermediate
CM-Sephadex	Carboxymethyl	Dextran	Weak
SP-Sephadex	Sulfopropyl	Dextran	Strong

C. Affinity Matrices

Matrix	Group specificity
5'-AMP - agarose	Enzymes that have NAD^+ cofactor; ATP-dependent kinases
Benzamidine -Sepharose	Serine proteases
Boronic acid - agarose	Compounds with *cis*-diol groups: sugars, catecholamines, ribonucleotides, glycoproteins
Cibracron blue - agarose	Enzymes with nucleotide cofactors (dehydrogenases, kinases, DNA polymerases), serum albumin
Concanavalin A - agarose	Glycoproteins and glycolipids
Heparin - Sepharose	Nucleic acid-binding proteins, restriction endonucleases, lipoproteins
Iminodiacetate - agarose	Proteins with affinity for metal ions, serum proteins, interferons
Lentil-lectin - Sepharose	Detergent-soluble membrane proteins
Lysine - Sepharose	Nucleic acids
Octyl - Sepharose	Weakly hydrophobic proteins, membrane proteins
Phenyl - Sepharose	Strongly hydrophobic proteins
Poly (A) - agarose	Nucleic acids containing poly (U) sequences, mRNA-binding proteins
Poly (U) - agarose	Nucleic acids containing poly (A) sequences, poly (U)-binding proteins
Protein A - agarose	IgG-type antibodies
Thiopropyl - Sepharose	-SH containing proteins

V. The Genetic Code

First position 5′-end	Second position				Third position 3′-end
	U	C	A	G	
U	Phe	Ser	Tyr	Cys	**U**
	Phe	Ser	Tyr	Cys	**C**
	Leu	Ser	*Stop	*Stop	**A**
	Leu	Ser	*Stop	Trp	**G**

C	Leu	Pro	His	Arg	**U**
	Leu	Pro	His	Arg	**C**
	Leu	Pro	Gln	Arg	**A**
	Leu	Pro	Gln	Arg	**G**
A	Ile	Thr	Asn	Asn	**U**
	Ile	Thr	Asn	Asn	**C**
	Ile	Thr	Lys	Lys	**A**
	Met	Thr	Lys	Lys	**G**
G	Val	Ala	Asp	Gly	**U**
	Val	Ala	Asp	Gly	**C**
	Val	Ala	Glu	Gly	**A**
	Val	Ala	Glu	Gly	**G**

VI. Enzymes in Molecular Cloning

Alkaline Phosphatase

Removes phosphate from 5′-ends of ds or ss DNA or RNA

DNA Ligase (Phage T_4)

Joins sugar-phosphate backbones of ds DNA with a 5′-P and a 3′-OH in an ATP-dependent reaction. Requires that the ends of the DNA be compatible.

DNA Polymerase I

Synthesiszes DNA complementary to a DNA template in a 5′ → 3′ direction beginning with a primer with a free 3′-OH. The Klenow fragment is a truncated version of DNA polymerase I which lacks the 5′ → 3′ exonuclease activity.

Exonuclease III

Exonucleases cleave from the ends of linear DNA. Exonuclease III digests ds DNA from the 3′-end only.

Mung Bean Nuclease

Digests ss nucleic acids but will leave intact any region which is double helical.

Nuclease S1

Digests ss nucleic acids but will leave intact any region which is double helical. However, the enzyme will also cleave a strand opposite a nick on the complementary strand.

Polynucleotide Kinase

Adds phosphate to 5′-OH ends of ds or ss DNA or RNA on an ATP-dependent reaction. If [U-^{32}P] ATP is used, then the DNA will become radioactively labeled.

Restriction Enzymes

Cut both strands of ds DNA within a (normally symmetrical) recognition sequence. Hydrolyse sugar-phosphate backbone to give a 5′-P on one side and a 3′-OH on the other. Yield blunt or sticky ends (5′ or 3′ overhangs).

Reverse Transcriptase

RNA-dependent DNA polymerase. Synthesizes a complementary DNA strand (cDNA) from an RNA template in a 5′ → 3′ direction beginning with a primer with a free 3′-OH. Requires dNTPs.

RNase A

Nuclease which digests RNA, but not DNA

RNase H

Nuclease which digests the RNA strand of an RNA-DNA hybrid.

T7, T3 and SP6 RNA Polymerases

Specific RNA polymerases encoded by the respective bacteriophages. Each enzyme recognizes only the promoters from its own phage DNA, and can be used specifically to transcribe DNA downstream of such a promoter.

Taq DNA Polymerase

DNA polymerase derived from a thermostable bacterium, *Thermus aquaticus*. Operates at 72°C and is reasonably stable above 90°C. Used in PCR.

Terminal Transferase

Adds a number of nucleotides to the 3′-end of linear ss or ds DNA or RNA. If only GTP is used for example, only Gs will be added.

VII. Restriction Enzymes

A. Diluent Buffers for Restriction Enzymes

Diluent buffers are required for dilution of restriction enzyme prior to performing a restriction digestion. The diluted enzyme should be used the same day. GENEI provides prepared diluent buffers appropriate for use in a restriction assay and provides a code of tips for the use of restriction enzymes.

The composition of the diluent buffers is:

Diluent buffer	Constituents	Type of restriction enzyme
1	10 mM Tris.HCl, pH 7.4 50 mM KCl 0.1 mM EDTA 1 mM DTT 200 μg/ml nuclease free BSA 50% glycerol	*Acc* I, *Apa*L I, *Ava* I; *Bcl* I, *Bst*E II, *Cla* I, *Dra* I, *Eco*R V, *Hae* III, *Hha* I, *Hinc* II, *Hinf* I, *Hpa* I, *Hpa* I, *Hpa* II, *Kpn* I, *Mbo* I, *Mlu* I, *Msp* I, *Nae* I, *Nar* I, *Nco* I, *Nhe* I, *Nru* I, *Nsi* I, *Pvu* II, *Sal* I, *Sau*3A I, *Sau*96 I, *Sma* I, *Ssp* I, *Spe* I, *Stu* I, *Xho* I, *Xma* I, *Apa* I
2	10 mM Tris.HCl, pH 7.4 300 mM NaCl 0.1 mM EDTA 1 mM DTT 500 μg/ml nuclease free BSA 50% glycerol	*Alu* I, *Bgl* I, *Bgl* II, *Hind* III, *Pvu* I, *Sac* I, *Taq* I
3	10 mM Tris.HCl, pH 7.4 250 mM NaCl 0.1 mM EDTA 1 mM DTT 200 μg/ml nuclease free BSA 0.15% Triton X-100 50% glycerol	*Bam*H I, *Eco*R I, *Not* I, *Pst* I, *Sfi* I, *Xba* I

B. Tips to Use Restriction Enzymes

Some of the essential tips to use a restriction enzyme are:

(1) All restriction enzymes and 10X assay buffers should be stored at -20°C. Restriction enzymes should be kept on ice when they are not in the freezer. Thaw the assay buffers completely before using.
(2) The enzyme should be the last component added to the reaction mixture.
(3) Prior to a restriction digest, the substrate DNA should not be stored in TE buffer, since EDTA would complex Mg^{2+} ions required for catalysis by the restriction enzyme.
(4) The restriction enzyme : DNA : reaction volume ratio according to the unit definition is, 1unit : 1µg : 50 µl and can be used as a guide when designing a reaction mixture.
(5) Proper mixing to ensure complete digestion is required.
(6) A short spin after pipetting the enzyme would be recommended.
(7) Optimal pH for restriction enzymes ranges from 7.2–7.8.
(8) Optimal temperatures for the reaction varies from 30°C to 70°C, but most enzymes function at 37°C.
(9) Addition of BSA to the reaction mixture stabilizes some restriction enzymes, especially during longer incubations.
(10) The reaction can be terminated by heat inactivation at 65°C for 20 min. or addition of EDTA or by phenol/chloroform extraction (since some restriction enzymes may not be heat-inactivated). An alternate method of spin column may be used in case of any inactivation of DNA by heat (Methods in Enzymology, Recombinant DNA, Part G, 1992).

VIII. Nucleic Acid Conversions

A. Spectrophotometric Conversions

A_{260} nm	DNA	RNA	Concentration	
			µg/ml	mM (Nucleotides)
1.0	ds	—	50	0.15
1.0	ss	—	33	0.1
1.0	—	ss	40	0.12
6.7	ds	—	—	1.0
10.0	ss	—	—	1.0
8.3	—	ss	—	1.0

B. Nucleotide MWs

Average MW of deoxyribonucleotide base (dNTP) = 333 Da
Average MW of ribonucleotide base (NTP) = 340 Da

C. Molar Conversions

1 μg of 1000 bp DNA	=	1.52 pmol
1 μg of pUC18/19 DNA (2686 bp)	=	0.35 pmol
1 μg pBR322 DNA (4361 bp)	=	0.29 pmol
1 μg SV40 DNA (5243 bp)	=	0.29 pmol
1 μg φX 174 DNA (5386 bp)	=	0.28 pmol
1 μg M13mp18/19 DNA (7250 bp)	=	0.21 pmol
1 μg of λ phage DNA (48502 bp)	=	0.03 pmol
1 pmol of 1000 bp DNA	=	0.66 μg
1 pmol of pUC18/19 DNA (2686 bp)	=	1.77 μg
1 pmol pBR322 DNA (4361 bp)	=	2.88 μg
1 pmol SV40 DNA (5243 bp)	=	3.46 μg
1 pmol φX 174 DNA (5386 bp)	=	3.54 μg
1 pmol M13mp18/19 DNA (7250 bp)	=	4.78 μg
1 pmol of λ phage DNA (48502 bp)	=	32.01 μg

IX. Nucleotide and Protein Databases

A. Primary Nucleotide Sequence Databases

EMBL (European Molecular Biology Laboratory)
http://www.ebi.ac.uk/embl/
GENBANK @ NCBI (National Center for Biological Information)
http://www.ncbi.nlm.nih.gov
DDBJ (DNA Databank of Japan)
http://www.ddbj.nig.ac.jp

B. Secondary Nucleotide Sequence Databases

UNIGENE
http://www.ncbi.nlm.nih.gov/Unigene/
SGD Genome (*Saccharomyces* Genome Database)
http://www.yeastgenome.org/
EBI Genomes
http://www.ebi.ac.uk/genomes/
Ensembl
http://www.ensembl.org/index.html

C. Protein Sequence Databases

SWISS-PROT, TrEMBL
http://www.expasy.ch/sprot/

Developed by SWISS-PROT and TrEMBL groups at Swiss Institute of Bioinformatics and at EBI
PIR (Protein Information Resource)
http://pir.georgetown.edu
Division of National Biomedical Research Foundation (NBRF), USA in collaboration with Munich Information Center for Protein Sequences (MIPS) and Japanese International Protein Sequence Database (JIPID)

D. Sequence Motif Databases

Pfam (database of protein families defined as domains)
http://pfam.sanger.ac.uk/
PROSITE
http://www.expasy.ch/prosite

E. Macromolecular 3-D structure Database

PDB (Protein DataBank)
http://www.rcsb.org/pdb
SCOP (Structural Classification of Proteins)
http://scop.mrc-lmb.cam.ac.uk/scop/
CATH (Class, Architecture, Topology, Homologous superfamily)
http://www.biochem.ucl.ac.uk/bsm/cath

F. General Databases

GeneCards (database of human genomes)
http://www.genecards.org/
KEGG (Kyoto Encyclopedia of Genes and Genomes)
http://www.genome.ad.jp/kegg/

X. GENBANK FORMAT FOR NUCLEOTIDE SEQUENCES

A. GenBank is an open official database for deposit of Nucleotide sequences in the NCBI server (http://www.ncbi.nlm.nih.gov), where nucleotide sequences can also be retrieved by identification of its accession numbers or by keyword search in the NCBI homepage. A GenBank Nucleotide sequence deposition implies that the researcher can publicly declare his unique contribution of work upon submission and make his sequence freely available to other investigators. A number of sequences are constantly spun in several laboratories and there has been a marked rise in sequence deposits. Several other databases exist, but GenBank is a widely used standard database. A familiarity with GenBank format is essential to know the attributes of any Nucleotide sequence.

Consider the following GenBank format for *Candida rugosa* lipase 5 gene that has been deposited in 2006. The features are explained herewith to allow the reader to analyse the sequence from its attributes:

B. GenBank Nucleotide sequence for *Candida rugosa* lipase 5 gene with Accession ID DQ984523. *Candida rugosa* lipase [gi:119091223]

Features	Sequence
LOCUS	DQ984523 1625 bp DNA linear PLN 28-JUL-2008
DEFINITION	*Candida rugosa* lipase 5 gene, promoter region and 5' UTR.
ACCESSION	DQ984523
VERSION	DQ984523.1 GI:119091223
KEYWORDS	
SOURCE	*Candida rugosa*
ORGANISM	*Candida rugosa* Eukaryota; Fungi; Dikarya; Ascomycota; Saccharomycotina; Saccharomycetes; Saccharomycetales; mitosporic Saccharomycetales; Candida
REFERENCE	1 (bases 1 to 1625)
AUTHORS	Hsu,K.H., Lee,G.C. and Shaw,J.F.
TITLE	Promoter analysis and differential expression of the *Candida rugosa* lipase gene family in response to culture conditions
JOURNAL	J. Agric. Food Chem. 56 (6), 1992-1998 (2008)
PUBMED	18290622
REFERENCE	2 (bases 1 to 1625)
UTHORS	Hsu,K.-H., Lee,G.-C. and Shaw,J.-F.
TITLE	Direct Submission
JOURNAL	Submitted (05-SEP-2006) Department of Food Science and Biotechnology, 250 Kuo Kuang Rd., Taichung, Taiwan 402, R.O.C.
FEATURES	Location/Qualifiers
source	1..1625 /organism="Candida rugosa" /mol_type="genomic DNA" /db_xref="taxon:*5481"*
gene	1..>1625 /gene="lipase 5"
misc_feature	1..1625 /gene="lipase 5" /note="contains promoter and 5' UTR"

ORIGIN

1 aacgacgtga ggttgacgct gtagctgcga agctcctcca caaatggata ttccttttca

61 agcatgtcca acctcaacgc aaggttgtcc acatgctcga tgtagtcggt ccatgctccc
121 caggtcggat gggtgaaata aaggaaggca aactcgagcg tgtgcacacg gtacgggaag
181 gggagtctgg cctcggctct ggctgtaaaa ctggaccatt catagtatcc cgccactagg
241 tcagcacttc cacggactgc aaactccagc cactgtcgcc tgcgagcaag gtctcggatc
301 tcaggcacgt ggcaattccg gtagagttcg gcggtgctct tgggaaggtg cacaaacaaa
361 gttgcaagca cgaggtacgg aagcccgcac acgtgctcca caatgtcaaa gctcatgggg
421 aataatgtgg cggcggagct acggcgataa ggtcatccac ggttatgcat acaagaggca
481 gaactgcttg tgtgcaggac agcagggtgc cgcgttcact ggcggtgagg tcgttgacat
541 taccgtcaac cacgttgatt ttgtcaaggg caacccggta aaatgaccga gtgcttgcct
601 cagaatggta gcagtgcggg agttattcgg gtggatgcaa accctcaggc tcttcaaccc
661 cgcaggtagg aaaggtgcca atggagaaat gtccacaata gggaacctga aaagcctgag
721 cctgagcaac ccctctgcga gctgcggtat ttggtccgca tactgtgaac ggatatggga
781 gagaaagctc ggtttctccc tccatccggg tgtgcatcag attacatccc ataatccact
841 cctttcgcgt tcaccccac agagacacga agcacctccc cactccccat tcacaaaccc
901 tcgtctatgg aagacatgca tgaggcgtgg ttgcaccaga ccctaatgat catcacactg
961 gcttgccgcg cccaatcccc ggccatcgtc cattcctgaa acaaacaagc aacgttgacg
1021 caggtatgca gacacccact gctgtatacg ccctgtccgt cgtgcttctt ttcagatccc
1081 cgtccacacc gtgtgcattt tttgctggtg ttggcaacgg atacaaagcc atggggagtc
1141 aacaaacccc ccaaacactc cgggttctat caggcacacg caaacggcat ttgcaccccc
1201 tttccgatga tccgcacatg tggacccgcc accaagcctg caggaccatt aacggcggag
1261 atcactgcac cgcagacgag catgtatccg cccgttccgc agcgtgctct ggataaaggt
1321 actcaaagtc gacagaatct cagccaggtg gcgtacagag aatagccata cccaccaagc
1381 aatgcccacc aagccatgtc ccaccaagtc acatttggac acaaacgcag tgccagggtt
1441 tcacaactgg ctcctacctg acgcatctcg gcaaatgcca ctctgtagtt ggtccagaca
1501 tggcaggcat aaacaagcac caggcggtca gacacaagag cgtggggagg tgacaatgcc
1561 gatcgcggac cagtttacac cgctccgagt ataaaagcag aagcattctc acctgctcgc
1621 tcccc
//

C. Interpretation of GenBank Sequence Features

LOCUS Accession number: DQ984523
Number of bases: 1625 bp
Section of the database: DNA linear
Date of submission: PLN 28-JUL-2008
(Date of submission of sequence: 05-Sep-2006)

DEFINITION A concise description of the sequence

ACCESSION Accession number, a constant code assigned to each entry DQ984523

NID Nucleotide identifier, GI:119091223, a unique reference to the current version of the sequence information; this allows the sequence to be revised, but still be associated with the same locus name and the accession number.

KEYWORDS Introduces a list of short phrases assigned by the author, describing gene products and other relevant information

	about the entry
SOURCE	Information on organism from which data is derived
ORGANISM	Biological classification of the source organism
REFERENCE	Portion of sequence data to which the cited literature refers- AUTHOR, TITLE, JOURNAL, PUBMED ID
FEATURES	Describes properties of sequence in detail
SOURCE	Source of organism; molecular type (genomic DNA); link to a taxonomic database (db_xref="taxon:5481)
GENE	Coding sequence (CDS) of the gene lipase 5, from bases 1-1625
MISC	A note on the gene sequence attributes, in this case, the CDS contains promoter and 5'-UTR (untranslated region)
ORIGIN	location of the first base of the sequence within the gene
//	Marker for termination of the entry

XI. BIOINFORMATICS TOOLS FOR PROTEIN STRUCTURE PREDICTION

A. Protein structure prediction is an important area of protein science. Every protein has a primary structure, its sequence, a secondary structure, the helices and β-sheets and a tertiary structure, the folding of the protein in a proper conformation. Protein structure has been experimented for the past several decades by physical and chemical methods of investigation. With the advent of bioinformatics, protein structure has been aimed to be predicted from the knowledge of its primary sequence. A number of servers are available online for prediction of the protein parameters, patterns and profiles, primary, secondary and tertiary structures (ExPASY) and protein visulalization (Rasmol). A number of methods exist for protein structure prediction such as Threading, *Ab initio* prediction and comparative modeling.

B. Homology Modeling

Name	Method
CABS	Reduced modeling tool
MODELLER	Satisfaction of spatial restraints
ROSETTA	Rosetta homology modeling
SWISS-MODEL	Local similarity/fragment assembly
TIP-STRUCTFAST	Automated Comparative Modeling
WHAT IF	Position specific rotamers

C. Threading/fold Recognition

Name	Method
PSI-BLAST	Iterative sequence alignment for fold identification
3D-PSSM	3D-1D sequence profiling
SUPERFAMILY	Hidden Markov modeling
mGenTHREADER/GenTHREADER	Sequence profile and predicted secondary structure
LOOPP	Multiple methods

D. Ab Initio Structure Prediction

Name	Method
ROSETTA	Rosetta homology modeling and *ab initio* fragment assembly
Rosetta@Home	Distributed-computing implementation of Rosetta algorithm
CABS	Reduced modeling tool

E. Secondary Structure Prediction

Name	Method
Jpred	Neural network assignment
PREDATOR	Knowledge-based database comparison
PredictProtein	Profile-based neural network

XII. Related Journals

Applied Biochemistry and Biotechnology
BBA – Gene structure and Expression
BBA – General Subjects
BBA – Proteins and Proteomics
Biochemical Journal
Biochemistry
Biochimie
Bioscience, Biotechnology and Biochemistry
Biotechniques
Biotechnology letters
Biotechnology Progress
Electronic Journal of Biotechnology
EMBO Journal
EMBO reports
Journal of Biotechnology
Journal of Molecular Biology
Microbiology
Molecular Biotechnology
Nature Biotechnology
Nature Methods
Nature Protocols
Protein expression and Purification
Protein Science
Trends in Biochemical Sciences
Trends in Biotechnology

Section III

ABBREVIATIONS

A	Adenine
A_{280}	Absorbance at 280 nm
A_{260}	Absorbance at 260 nm
A_{500}	Absorbance at 500 nm
A_{715}	Absorbance at 715 nm
Ab^r	Antibiotic resistance
Amp^r	Ampicillin resistance
A^o	Angstrom = $1x10^{-1}$ nm
APS	Ammonium persulfate
Arg	Arginine
Asp	Aspartic acid
ATP	Adenosine triphosphate
ATPS	Aqueous two phase system
BAC	Bacterial artificial chromosome
BCA	Bicinchoninic acid
BLAST	Basic Local Alignment Search Tool
Bp	Base pair
BSA	Bovine serum albumin
C –	Carbon source
C	Carbon / Cytosine
Ca	Calcium
$CaCl_2$. $2H_2O$	Calcium chloride dihydrate
CBB	Coomassie Brilliant Blue
CD	Circular dichroism
CDA	Czapek-Dox agar medium
cDNA	Complementary DNA
cfu	colony forming units
cm	centimetre
CO_2	Carbon dioxide
Con A	Concanavalin A
CRL	*Candida rugosa* lipase
CTAB	Cetyl Trimethyl Ammonium Bromide

C-terminus	Carboxyl terminus
$CuSO_4.5H_2O$	Copper sulfate pentahydrate
CVL	*Chromobacterium viscosum* lipase
DAG	Diacylglycerol
DDBJ	DNA Database of Japan
DIPF	Diisopropylfluorophosphate
DMF	Dimethyl fluoride
DMSO	Dimethyl sulfoxide
DNA	Deoxyribonucleic acid
dNTP	Deoxyribonucleotide triphosphate
DOE	Department of Energy, USA
dTMP	Deoxythymine monophosphate
DTT	Dithiothreitol
E	Enantioselectivity
E.coli	*Escherichia coli*
E_a	Arrhenius energy of activation
EBI	European Bioinformtics Institute
EDTA	Ethylene diamine tetraacetate
ELISA	Enzyme-linked Immunosorbent assay
EMBL	European Molecular Biology Laboratory
EtBr	Ethidium bromide
Fe	Iron
$FeCl_2$	Ferrous chloride
$FeCl_3$	Ferric chloride
$FeSO_4$ $(NH_4)_2SO_4.6H_2O$	Ferrous ammonium sulfate hexahydrate
$FeSO_4$ $.7H_2O$	Ferrous sulfate heptahydrate
FFA	Free fatty acid
FPLC	Fast protein liquid chromatography
G	Glycerol / Glycine / Guanine
g	Gram
g/L	Gram per liter
GC	Gas chromatography
GdmCl	Guanidium chloride
GI tract	Gastrointestinal tract
Glu	Glutamic acid
GMO	Genetically modified organism
GMP	Good manufacturing practice
GRAS	Generally regarded as safe
GSB	Glycerol storage buffer
GST	Glutathione-S-transferase
h	Hour
H-bonding	Hydrogen bonding
HCl	Hydrochloric acid
HEPES	4-(2-hydroxyethyl)-1-piperazine ethanesulfonic acid
HGP	Human Genome Project
HIC	Hydrophobic interaction chromatography

His	Histidine
HLL	*Humicola lanuginosa* lipase
HPLC	High pressure liquid chromatography
HTTP	HyperText Transport Protocol
IAA	Indole acetic acid
IEF	Isoelectric focusing
IP	Internet protocol
IPTG	Isopropyl β-D-1-thiogalactopyranoside
IR	Infra red spectroscopy
JIPID	International Protein Information database of Japan
K_2HPO_4	Dipotassium hydrogen orthophosphate
Kb	Kilobase
KCl	Potassium chloride
kDa	KiloDaltons
KH_2PO_4	Potassium dihydrogen orthophosphate
L	Liter
LB	Luria-Betani agar medium
LED	Lipase Engineering database
µg	microgram
µl	microliter
M	Molar concentration
MAG	Monoacylglycerol
MCS	Multiple cloning site
MD	Maryland, USA
MES	2-(*N*-morpholino) ethanesulfonic acid
Mg	Magnesium
mg	Milligram
$MgSO_4 .7H_2O$	Magnesium sulfate heptahydrate
min	Minute
MIPS	Matinsried Institute for Protein Sequences
ml	Milliliter
Mn	Manganese
$MnCl_2$	Manganese chloride
$MnSO_4. 7H_2O$	Manganese sulfate heptahydrate
$MnSO_4.4H_2O$	Manganese sulfate tetrahydrate
Mo	Molybdenum
MOPS	3-(N-morpholino) propanesulfonic acid
ms	millisecond
MW	Molecular weight
N –	Nitrogen source
Na	Sodium
Na_2CO_3	Sodium carbonate
Na_2HPO_4	Disodium hydrogen phosphate
NaCl	Sodium chloride
NAD^+	Nicotinamide Adenine Dinucleotide
NaOH	Sodium hydroxide

NBRF	National Biomedical Research Foundation
NCBI	National center for Biotechnology Information
NCHGR	National Center for Human Genome Research, USA
ng	nanogram
NH_4Cl	Ammonium chloride
$(NH_4)_2HPO_4$	Diammonium hydrogen phosphate
$(NH_4)_2SO_4$	Ammonium sulfate
$(NH_4)_6Mo_7O_{24}.4H_2O$	Ammonium molybdate tetra hydrate
NHGRI	National Human Genome Research Institute, USA
Ni	Nickel
NIH	National Institutes of Health, USA
nm	nanometre=10^{-9} m
NMR	Nuclear magnetic resonance spectrosopy
O/N	Overnight
OD	Optical density
PAGE	Polyacrylamide gel electrophoresis
PCL	*Pseudomonas cepacia* lipase
PCR	Polymerase chain reaction
PDB	Protein data Bank
PEG	Polyethylene glycol
PEL	*Penicillium expansum* lipase
PERL	Practical Extraction and Reporting Language
PGL	*Pseudomonas glumae* lipase
Phe	Phenylalanine
PIPES	Piperazine-1,4-bis(2-ethanesulfonic acid)
pmol	Picomole=10^{-12} moles
PMSF	Phenyl methyl sulphonyl fluoride
p-NPP	para-Nitrophenyl phosphate
Ps. mendocina	*Pseudomonas mendocina*
PVA	Polyvinyl alcohol
PVDF	Polyvinylidene difluoride
R-, R1-, R2-, R3-	Alkyl groups
rDNA	Recombinant deoxyribonucleic acid
RmL	*Rhizomucor miehei* lipase
RNAi	RNA (Ribonucleic acid) interference
RNase	Ribonuclease
Rpm	rotations per minute
RT	Room temperature
RT-PCR	Reverse transcriptase polymerase chain reaction
S, Ser	Serine
SCP	Single cell protein
SDS	Sodium dodecyl sulfate
SIB	Swiss Institute of Bioinformatics
SmF	Submerged fermentation
SOD	Superoxide dismutase
sp.	species of the microbial taxonomic name

SRCD	Synchroton radiation circular dichroism
S-S bond	Disulfide bond
ssDNA	single stranded DNA
SSF	Solid state fermentation
T	Thymine
Ta	Annealing temperature
TAG	Triacylglycerol
Taq	*Thermus aquaticus*
TCA	Trichloroacetic acid
TCP	Transmission control protocol
TE	Tris-EDTA buffer
TEMED	N,N,N′,N′-Tetramethylethylenediamine
Tet^r	Tetracycline resistance
Thiamine.HCl	Thiamine hydrochloride
Thr40	Threonine at amino acid position 40
TLC	Thin layer chromatography
Tm	Melting temperature of DNA
Tris	tris (hydroxymethyl) aminomethane
Trp	Tryptophan
Tyr	Tyrosine
U	Units of enzyme activity
URL	Uniform Resource locator
UV	ultra-violet
v/v	Volume of solute per volume of solvent
Val	Valine
X	Any amino acid
X-gal	5-Bromo-4-chloro-Indoly-β-D-Galactoside
XML	Extensible Markup language
YAC	Yeast artificial chromosome
Zn	Zinc
$ZnSO_4.7H_2O$	Zinc sulfate heptahydrate
β ME	beta-mercaptoethanol
λ	wavelength
2-D NMR	2-dimensional NMR spectroscopy
2-D PAGE	2-dimensional polyacrylamide gel electrophoresis
3-D	Three-dimensional

Section IV

REFERENCES

I. INTERNET

http://www.wikipedia.org
http://www.genome.gov
http://www.promega.com
http://www.stratagene.com
http://www.images.google.com
http://www.mcb.uct.ac.za/pcroptim.htm
http://www.promega.com/biomath/calc11.htm

II. BOOKS

Atlas RM (1993) *Handbook of Microbiological Media*, CRC Press.

Attwood TK & Parry-Smith DJ (2003) *Introduction to Bioinformatics*, Pearson Education, New Delhi.

Boyer, R (2001) *Modern Experimental Biochemistry*, Third Edition, Addison Wesley Longman, New Delhi.

Harris ELV & Angal S (1989) *Protein purification methods- a practical approach*, Oxford University Press, New York.

Sambrook J & Maniatis T (1989) *Molecular cloning*: A Laboratory Manual, Second Edition, Cold Spring Harbor Laboratory Press, New York.

Turner PC, McLennan AG, Bates AD & White MRH (2001) *Instant Notes Molecular Biology*, Second Edition, Viva Books Private Limited, New Delhi.

Westhead DR, Parish JH & Twyman RM (2003) *Instant Notes Bioinformatics*, Viva Books Private Limited, New Delhi.

III. Journal Articles

Bachinsky, AG; Frolov, AS; Naumochkin, AN et al. (2000) PROF_PAT 1.3: Updated database of patterns used to detect local similarities. *Bioinform*, 16: 358-366.

Basheer, S; Mogi, K & Nakajima, M (1995) Surfactant-modified lipase for the catalysis of the interesterification of triglycerides and fatty acids. *Biotechnol Bioeng*, 45: 187-195.

Bian, CB; Yuan, C; Lin, L et al. (2005) Purification and preliminary crystallographic analysis of a *Penicillium expansum* lipase. *Biochim Biophys Acta*, 1752: 99-102.

Bornscheuer, UT (2002) Methods to increase enantioselectivity of lipases and esterases. *Curr Opin Biotechnol*, 13: 543-547.

Chung, CT; Niemela, SL & Miller, RH (1989) One-step preparation of competent *Escherichia coli*: transformation and storage of bacterial cells in the same solution. *PNAS*, 86: 2172-2175.

Del Vecchio, P; Graziano, G; Granata, V et al. (2002) Denaturing action of urea and guanidine hydrochloride towards two thermophilic esterases. *Biochem J*, 367: 857-863.

Derewenda, ZS; Derewenda, U & Dodson, GG (1992) The crystal and molecular structure of the *Rhizomucor miehei* triacylglyceride lipase at 1.9A^{0} resolution. *J Mol Biol*, 227: 818-839.

Gracy, J & Argos, P (1998) DOMO: a new database of aligned protein domains. *TIBS*, 23: 495-497.

Jutila, A; Zhu, K; Tuominen, EKJ et al. (2004) Fluorescence spectroscopic characterization of *Humicola lanuginosa* lipase dissolved in its substrate. *Biochim Biophys Acta*, 1702: 181-189.

Kamiya, N; Goto, M & Nakashio, F (1995) Surfactant-coated lipase suitable for the enzymatic resolution of menthol as a biocatalyst in organic media. *Biotechnol Prog*, 11: 270-275.

Kelly, SM; Jess, TJ & Price, NC (2005) How to study proteins by circular dichroism. *Biochim Biophys Acta*, 1751, 119-139.

Kohno, M; Kugimiya, W; Hashimoto, Y et al. (1993) Preliminary investigation of crystals of lipase I from *Rhizopus niveus*. *J Mol Biol*, 229: 785-786.

Ladner, CL; Yang, J; Turner, RJ et al. (2004) Visible fluorescent detection of proteins in polyacrylamide gels without staining. *Anal Biochem*, 326: 13-20.

Liebeton, K; Zacharias, A & Jaeger, KE (2001) Disulfide bond in *Pseudomonas aeruginosa* lipase stabilizes the structure but is not required for interaction with its foldase. *J Bacteriol*, 183: 597-603.

Mala, JGS & Takeuchi, S (2008) Understanding structural features of microbial lipases-an overview. *Anal Chem Insights*, 3: 9-19.

Melo, EP; Taipa, MA; Castellar, MR et al. (2000) A spectroscopic analysis of thermal stability of the *Chromobacterium viscosum* lipase. *Biophys Chem*, 87: 111-120.

Miled, N; Riviere, M; Cavalier, JF et al. (2005) Discrimination between closed and open forms of lipases using electrophoretic techniques. *Anal Biochem*, 338: 171-178.

Peterson, MTN; Fojan, P & Peterson, FB (2001) How do lipases and esterases work: the electrostatic contribution. *J Biotechnol*, 85: 115-147.

Pleiss, J; Fischer, M; Peiker, M et al. (2000) Lipase engineering database. Understanding and exploiting sequence-structure-function relationships. *J Mol Catal B Enz*, 10: 491-508.

Rahman, RNZRA; Baharum, SN; Basri M et al. (2005) High-yield purification of an organic solvent-tolerant lipase from *Pseudomonas* sp. strain S5. *Anal Biochem*, 341: 267-274.

Ransac, S; Blaauw, M & Dijkstra, B (1995) Crystallization and preliminary X-ray analysis of a lipase from *Staphylococcus hyicus*. *J Struct Biol*, 114: 153-155.

Retamal, CA; Thiebaut, P & Alves, EW (1999) Protein purification from polyacrylamide gels by sonication extraction. *Anal Biochem*, 268: 15-20.

Saul, A & Don, M (1984) A rapid method of concentrating proteins in small volumes with high recovery using Sephadex G-25. *Anal Biochem*, 138: 451-453.

Schrag, JD; Li, Y; Cygler, M et al. (1997) The open conformation of a *Pseudomonas* lipase. *Structure*, 5: 187-202.

Sheen, H & Ali-Khan, Z (2005) Protein sample concentration by repeated loading onto SDS-PAGE. *Anal Biochem*, 343: 338-340.

Shin, M; Sakihama, N; Oshino, R et al. (1984) Butyl-Toyopearl 650 as a new hydrophobic adsorbent for water-soluble enzyme proteins. *Anal Biochem*, 138: 259-261.

Svendsen, A (2000) Lipase protein engineering. *Biochim Biophys Acta*, 1543: 223-238.

Wessel, D & Flugge, UI (1984) A method for quantitative recovery of protein in dilute solution in the presence of detergents and lipids. *Anal Biochem*, 138: 141-143.

Zhang, J & Greasham, R (1999) Chemically defined media for commercial fermentations. *Appl Micrbiol Biotechnol*, 51: 407-421.

INDEX

C

T

U

V